First published in 2011 by New Holland Publishers (UK) Ltd
London • Cape Town • Sydney • Auckland
Garfield House, 86–88 Edgware Road, London W2 2EA, United Kingdom
www.newhollandpublishers.com
80 McKenzie Street, Cape Town 8001, South Africa
Unit 1, 66 Gibbes Stree, Chatswood, NSW 2067, Australia
218 Lake Road, Northcote, Auckland

ISBN 978-184773-748-9
10 9 8 7 6 5 4 3 2 1
Publisher: Clare Sayer Senior Editor: Lisa John Production: Laurence Poos
Research Advisor: Patricia H. Squire MA
Designed and created for New Holland by AG&G Books Copyright © 2004 "Specialist" AG&G Books
Design: Glyn Bridgewater and Ginny Zeal Illustrations: Coral Mula
Editor: Alison Copland Photographs: iStockphoto
Reproduction by Pica Digital Pte Ltd, Singapore
Printed and bound in Malaysia by Times Offset (M) Sdn. Bhd.
The information in this book is true and complete to the best of our knowledge. All recommendations
are made without guarantee on the part of the authors and the publishers. The authors and publishers
disclaim any liability for damages or injury resulting from the use of this information.

The **HENKEEPING**

... guide to choosing ...d keeping chickens for egg and meat production

David Squire
Series editors: A. & G. Bridgewater

Contents

Author's foreword **2**

Author's foreword

Keeping chickens or other poultry in your back garden or yard becomes a way of life and one that is packed with the amusement provided by your dedicated team of egg layers. Some will bustle around as if parading their management skills, a few will continually inspect the ground for reckless insects, and others will exhibit a measured walk, the product of carefully cultivated self-importance. Of course, there is also the daily bonus of eggs for your family's breakfast or use in numerous kitchen recipes. To top it all, you will have the reassurance that your birds have led a happy, relaxed life within a near-natural environment and that they have been fed on a diet of wholesome food.

Keeping poultry, either in a wire-netted enclosure in your garden or as free-rangers if you have more space, is part of getting back to the basics of life and an activity that puts you into daily contact with the land and its seasons. Additionally, when living in harmony with the countryside, the cycle of life becomes clear. All poultry enthusiasts – and especially their children – are captivated by chicks breaking free from their shells and cautiously entering into a new world.

There are, of course, disadvantages to keeping poultry, and you will need to be constantly vigilant to prevent predators, such as foxes and rats, attacking your birds, but this problem need not be insurmountable if the hen house is strong and properly constructed, and the fences are high. Essentially, you should not enter the world of keeping poultry if you cannot offer a seven days a week commitment throughout the year. The daily tasks of feeding, watering and collecting eggs must not be neglected. Remember that happy hens lay the best and most eggs.

MEASUREMENTS

Both metric and imperial measurements are given in this book. For example, 1.8 m (6 ft).

The world of poultry

The range of poultry species kept for the eggs they produce and the meat they provide is wide. This mainly comprises chickens, with ducks, guinea fowl and quails (see pages 68–73) playing a lesser role. However, like chickens, they have an enthusiastic and dedicated following. The number of chickens kept for their eggs or meat far exceeds that of other poultry, with a current estimated global figure of more than 28 billion, which is expected to rise.

What does 'poultry' mean?

WHAT IS THE ORIGIN OF DOMESTICATED CHICKENS?

Chickens have been domesticated and kept for their eggs and meat for many thousands of years. Now scientifically known as *Gallus gallus domesticus*, chickens were earlier thought to have been first domesticated in India, but now Southeast Asia and probably Vietnam are considered to be more accurate, at about 10,000 years ago.

From India, domestication spread to Asia Minor, then to Greece about 7,000 years ago and subsequently to Egypt during the 18th Dynasty (1550–1292 BCE). Chickens are now known worldwide and widely kept for their eggs or meat.

What is the chicken's ancestry?

Previously, the domesticated chicken was believed to have descended from both the Red Junglefowl (*Gallus gallus*) and the Grey Junglefowl (*Gallus sonneratii*). However, modern genetics indicate that the Grey Junglefowl is the most likely ancestor.

DOMESTICATED FOWL

Large-fowl chickens

These are very popular and there are many breeds to choose from (see pages 10–22)

Guinea fowl

Seed-eating, ground-nesting, partridge-like birds, sometimes known as guinea hens (see page 73)

Quails

Classified as game birds and related to pheasants, they can fly (see page 72)

Bantams

Smaller than large-fowl chickens, they are popular as egg layers (see pages 23–27)

Ducks

Easier to keep than chickens, they are often more prolific egg layers (see pages 68–71)

Getting to know chickens

Identifying a basic chicken is easy as they are widely known and publicized, but there are many different breeds (see pages 10–22), as well as smaller forms known as bantams (see pages 23–27). The majority of chickens you will see are hens (females) and it is these that produce the eggs that are widely sold and eaten each year throughout the world. The male chicken is a cock or rooster (known as a cockerel until it is 12 months old).

HOW CAN I TELL A HEN FROM A COCK?

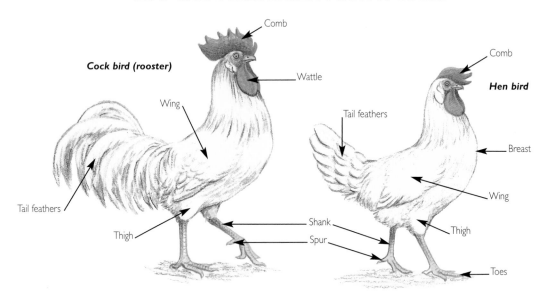

Cock bird (rooster) — Comb, Wattle, Wing, Tail feathers, Thigh, Shank, Spur

Hen bird — Comb, Tail feathers, Breast, Wing, Thigh, Toes

Hen chickens are usually smaller and weigh less than the male counterpart of their breed. In the breeds section of this book (pages 10–27), the weights of both hens and cocks are indicated. Additionally, a cock bird will have more flamboyant tail feathers and a larger and more colourful and distinctive comb on the top of his head. He also tends to strut and dominate the hens. It is the cock bird that makes the well-known 'cock-a-doodle-doo' noise that many people find disturbing, especially in urban areas.

The decision about whether it is a good idea to have a cock bird in with your hens is discussed on pages 50–51, and this depends on whether you just want to produce eggs or to breed from your chickens – or both. Most chicken enthusiasts, however, just keep hens.

Feather markings

Breeds are initially distinguished by their colours, shapes and sizes. A major part of this is their often spectacularly coloured feathers. Some breeds have just a single colour, while others have complicated patterning.

What is a broiler?

A term that originated in North America and is used to described a young bird, of either sex and under the age of eight weeks, specifically raised for its meat. The meat at that age is tender, with soft and pliable skin.

Psychology of a chicken

Chickens are inquisitive, intelligent and friendly and will continually keep you amused. Invariably they will come over and check what you are doing – so be careful you do not step on them. Bantam breeds are also reputed to be smarter than chickens, and they are certainly faster movers.

TYPES OF COMB

All chickens have a comb on their head. It is a fleshy growth and those on males are usually larger than ones on hens. The shapes and colours of combs vary from one breed to another. Many are red, but some are purple. Combs are not just present to give a chicken status and to make it more attractive to the opposite sex – they have a functional role in helping to cool a chicken during periods of high temperature. Incidentally, because blood is able to flow freely through a comb, it makes them extra sensitive during cold weather, especially if they are large. Coating a comb in petroleum jelly helps to provide cold-weather protection. There are several types of comb, including:

Rose
The comb lies almost flat on top and is broad, solid and very fleshy. The main part has a curved surface, with small, round protuberances. It ends in a single spike at the back of the head, called a leader.

Single
Thin and attached to the beak and along the skull. On the top there are 5–6 deep grooves that create several sharp points along the comb's top. This is the most common type of comb.

Strawberry
The comb is very low and is known as strawberry because of its shape. It has a rough surface.

Horn
Known also as the 'V-shaped comb', it has two pieces that look like horns, joined at their base.

Pea
Low, with three ridges. The middle ridge is slightly higher and larger than the other two.

Buttercup
The comb is cup shaped with regularly spaced points, resembling a crown. It is finely textured.

LEARNING THE PECKING ORDER – AND THIS MEANS YOU!

As you learn to live with chickens you will find that they put you as their leader – top in the pecking order! They will have decided that, as you provide their food, you will be revered. They will approach you in various ways, depending on their gender.

- **Cocks (also known as roosters):** The males will approach you cautiously, with direct eye contact and a slow flap of their wings.
- **Hens:** They might greet you in the same way as a cock bird, especially if they are feeling perky. This might be accompanied by the bird flapping her wings. If particularly friendly, she may also stand on your shoes and peck against your trousers. This is usually an approach that indicates she wants to be picked up. Unfortunately, this extra attention may create jealousy among the other hens.

Squabbling hens

Occasionally, squabbles occur between hens when they are not sure of the pecking order. Do not use your hands to separate them; rather, gently spray them with clean water. If further squabbles occur, it is usually best to rehome one of them.

Introducing new hens to an established group

This can be tricky but folklore suggests there are several ways to prevent trouble arising, including:

- Wait until night, when it is pitch dark.
- Rub garlic over the hen that is about to be introduced.

Keeping chickens

What is involved?

Before keeping chickens it is essential to understand what is involved in terms of time, initial and long-term cost, and the local and national regulations you will have to comply with. Below is a checklist of chicken-keeping questions to consider, ranging from whether it is worth the trouble and expense to what the ramifications of keeping chickens in close proximity to neighbours might be. Always get the goodwill of neighbours before keeping chickens.

CHICKEN CHECKLIST

Costs and results – is it worth it?

Do not expect to make a fortune out of keeping your own chickens or even to save a great deal of money on shopping bills – but you will be assured about the source of the eggs, their freshness, the foods which produced them and that the chickens have led a contented life. Eggs are one of the cleanest possible foods, the shell protecting the contents from dirt and germs.

Optimum number of chickens

Six hens will keep a family in eggs throughout the year. If you have a cock bird with them, this is about the number of hens that he can manage.

Eggs or meat?

Most backyard chicken enthusiasts keep hens solely for the eggs they produce. A list of the best egg-layers is given on page 8. Chickens that are raised for both meat and eggs are also listed on page 8.

Laying life of a hen

Pure-bred breeds can live for up to ten years (depending on the breed); this is much longer than hybrids, which often die before they reach five years of age. Most breeds start to lay eggs when 18–22 weeks old, but remember that the number of eggs decreases as a hen ages. Finally, it becomes uneconomic to keep her. Most hens will have reached the end of their economic life by the age of five.

Are chickens noisy?

Unless you have a cock bird in with your chickens (and it is not necessary to do this if all you want to do is to produce eggs), the noise level is not obtrusive. They will, however, make a fuss if alarmed and under attack. Also, some hens make a noise after laying an egg.

Can I keep just one chicken?

You can, but chickens are social creatures and need company. If you have just one chicken, it will be lonely and look to you to provide its social life! Keeping 3–6 chickens is about right to cater for their social activities.

Cock birds (roosters) are invariably noisy – but they are not essential for the production of eggs.

Rules and regulations

Keeping chickens on your land or allotment is surprisingly straightforward:

- First, check with the deeds of your property to ensure that you are legally able to keep chickens.
- Provided you have fewer than 50 chickens, there is no need to register them with any authority. Keeping more than 50 chickens is considered to be a commercial venture and must be registered with the authorities in your country.
- Consult with your local authority to check if there is a bye-law preventing anyone in your area from keeping chickens on their own property.
- Talk with your neighbours to see if they are in agreement with you keeping chickens; this is mainly a matter of getting their goodwill and to prevent disagreements arising later. Inviting them around to inspect your chickens and their housing usually settles any problems – especially if they are offered a few eggs!

CHICKEN CHECKLIST CONTINUED

Chickens are gregarious and therefore at their most contented when kept in small groups.

Chickens and household pets

Initially, cats and dogs will find chickens interesting, examining them as 'new boys on the block', but usually they soon lose interest. However, they need to be closely watched to check that they are not harming each other – much depends on the temperament of the dog or cat.

Chickens kept in wire-netting cages will not be at risk, but those foraging in a yard or living as free-rangers present a different problem.

Dogs: Hens that are chased and stressed by dogs will, at best, cease laying, or even be killed. If a dog kills a chicken, make it sit while you hold the dead chicken in front of it, at the same time scolding it in a harsh voice. Usually, young dogs soon learn to look and not touch, especially after receiving a peck on the nose from a dominant chicken. Occasionally, a dog will act as a protector, herding the chickens and barking to raise your attention if predators are a threat.

Cats: Initially inquisitive, they usually soon decide to respect and ignore each other. However, young chicks on their own are vulnerable unless accompanied by 'mum', who becomes a formidable opponent when her youngsters are under threat.

Trespassing chickens

Chickens that escape and wander into neighbouring gardens invariably cause problems. They eat plants, scratch and disturb soil and generally make a mess for which, in law, you will be responsible. If you take precautions to keep the hens securely fenced your liability could be less. Nevertheless, litigation can be expensive and time-consuming and it is better to regularly check your fencing and gates.

Incidentally, unlike geese and ducks, hens have no right of way on a road and if killed there is no obligation on the part of the motorist to pay compensation.

What about holidays?

If the holiday is a long day out, get up early to feed and water your chickens. However, if you are out all day and will not be back until late, either give them extra food or ask a neighbour to look after them. Automatic feeding and watering devices are invaluable in these circumstances.

When you are taking a holiday of several days, or longer, ensure that your chicken-sitter is a poultry enthusiast and has a basic knowledge of them and their routine. Write down details of feeding and watering (how much and when) so that there can be no mistakes.

Your own cat usually learns to live with the chickens, but neighbouring cats can be a problem.

Choosing the breed

What should I look for?

Before buying a bird, you need to decide if you want a chicken or a bantam (see page 10). Also think about whether your priority is for eggs rather than meat, or perhaps a combination of both (see below and page 39 for suitable breeds). Some chickens have a gentle and relaxed temperament that makes them ideal as pets for children – as well as for egg production – such as **Silkies**, both **Black** and **White** (see pages 12 and 22).

ARE THERE MANY DIFFERENT BREEDS?

There are hundreds of breeds of chickens and bantams, in many colours, sizes and egg-laying abilities. Some breeds have a long ancestry and are familiar to chicken enthusiasts and known as pure-breds, while others are more recent, such as hybrids. Each of these types has its own merits, and these are described on page 9.

For egg production

The presence of a male bird is essential for the production of chicks (see page 48) but not for the production of eggs.

Chickens are extremely gregarious and will be at their most contented when kept in a group.

LAYING OR TABLE BIRDS?

If space is limited to only a few chickens, most people choose breeds known for their egg-laying abilities, which is usually for several years and until keeping them becomes uneconomical; they are then killed (see pages 40–41). Those breeds kept for the table have a much shorter life than egg-laying ones. A few breeds are ideal for both egg-laying and the table.

Egg-laying breeds

- Barred Rock (see page 11)
- Black Australorp (see page 11)
- Cuckoo Maran (see page 15)
- Faverolles (see page 15)
- Leghorn (see page 18)
- Light Sussex (see page 18)
- Plymouth Rock (see page 20)
- Rhode Island Red (see page 21)
- Wyandotte (see pages 13, 17, 21 and 22)

Table breeds

- Croad Langshan (see page 15)
- Dorking (see pages 20, 21)
- Rhode Island Red (see page 21)

General-purpose breeds

- Barred Rock (see page 11)
- Black Australorp (see page 11)
- Buff Orpington (see page 14)
- Light Sussex (see page 18)
- Maran (see page 16)
- Plymouth Rock (see page 20)
- Rhode Island Red (see page 21)
- Wyandotte (see pages 13, 17, 21 and 22)

The White Silkie is a very distinctive and unusual-looking breed, with a docile and friendly nature.

TEN POPULAR BREEDS OF CHICKENS

The choice of chicken breeds is personal and differs from region to region, as well as from one country to another. Some breeds are selected for their egg-laying ability, others for their table qualities, while a few are chosen for their attractive appearance and character. Here are ten of the most popular breeds to consider:

- Araucana (see page 10)
- Barnevelder (see page 11)
- Black Rock (see page 12)
- Buff Orpington (see page 14)
- Red Dorking (see page 20)
- Light Sussex (see page 18)
- Rhode Island Red (see page 21)
- Silkie (see pages 12 and 22)
- Welsummer (see page 22)
- Wyandotte (see pages 13, 17, 21 and 22)

WHAT ARE PURE-BREDS?

These are breeds that have been popular for many years and when bred produce true replicas of themselves. They exhibit the distinct personality of the breed, unlike hybrids (see right) which are a combination of two breeds. There are many pure breeds to choose from, in a range of size, colour and plumage variations. Pure-bred chickens usually lay for more seasons than hybrid types, but are more likely to become broody, and consequently stop laying for a time. Also, some pure-bred chickens are difficult to source and can be expensive to buy. Nevertheless, for the home poultry keeper they are usually the breeds first considered when starting to keep chickens.

WHAT ARE HYBRID STRAINS?

These were first developed in the early 1950s in response to an increasing demand for eggs. They result when two pure-breds – such as Rhode Island Red, Marans and Leghorns – are crossed. Hybrids do not reliably pass on their features to their offspring (indeed, some hybrids are sterile), but are valued for the qualities they have acquired from their own parents.

Hybrid chickens are usually easy to look after, rarely go broody and reliably produce eggs. Unfortunately, egg-laying often decreases after their first season. Additionally, they can be prone to egg-laying problems and sometimes become aggressive in a flock of pure-breds.

A hen with a brood of chicks aimlessly strolling around her epitomizes the idyllic scenario of keeping chickens.

Chickens are expert scavengers, closely examining the ground in the hope of finding food.

Breeds of chicken

Are there many different breeds?

Chickens have been domesticated for hundreds of years, and during this time many different breeds have evolved. Most are known as pure-breds, and some as hybrids (see page 9). There is also a choice between chickens and bantams (see below). Here are descriptions of 37 popular breeds of chickens and seven of bantams. Always buy your birds from a reputable supplier, who will be able to provide you with healthy birds that are good examples of their breed.

CHICKEN OR BANTAM?

The range of both chickens and bantams is wide, and all of them are especially appealing when young and fluffy.

Chickens are larger than bantams and more popular with home poultry keepers as their eggs are larger and they provide more meat when kept for the table. Some bantam breeds have resulted from the selective crossing of chickens to create diminutive versions. There are also so-called 'true' bantams, which are naturally small and do not have large counterparts. Bantams are often raised for their ornamental qualities and many make excellent pets for children.

Because of their decreased size, bantams are often popular with poultry enthusiasts as they require less food and space than chickens. They also have a tendency to become broody and this assists when hatching eggs and raising further birds (see pages 50–53).

Araucana

Known in North America as South American Rumpless, this light, soft-feather breed has several specific qualities, including feathered ear tufts. Additionally, it is rumpless, having a complete absence of tail feathers. The American Poultry Club recognizes five colour variations (Black, White, Black-breasted Red, Silver Duckwing and Golden Duckwing), while the Poultry Club of Great Britain accepts four types (Large-tailed, Bantam-tailed, Large Rumpless and Bantam Rumpless).

Black Araucana

Temperament: Alert and active, strong, fast-growing chicks that mature quickly. Araucanas are subject to broodiness and make excellent mothers.

Eggs: 150–180 each year, sometimes more.

Egg colour: Blue, which permeates the entire shell.

Weight: Hen 2.3–2.7 kg (5–6 lb), cock 2.7–3.2 kg (6–7 lb).

Origin: In South America and named after the Arauca Indians in Northern Chile. In the early 1900s this breed was introduced into Europe. Incidentally, it should not be confused with the Ameraucana or Easter Egger, the latter a breed with imprecise parentage and variable traits.

Barnevelder

The Double-laced is the most popular variety of this heavy, soft-feather breed, with a rounded breast, broad shoulders and wings short and high. The head is neat, with a short, yellow beak revealing a dark point, orange eyes and yellow legs and feet. Body feathers have a red-brown ground colour, edged in black and with a shiny-green sheen. Neck hackles are black, also with a green sheen; those in the male have a red-brown edging.

Double-laced
Barnevelder

Temperament: Alert, well-balanced, good-natured and usually friendly with children. It tends to be lazy and therefore is best when in a free-range situation, where plenty of exercise is possible. Hens become broody fairly easily and make good mothers.

Eggs: 180–200 each year.

Egg colour: Brown.

Weight: Hen 2.72–3.2 kg (6–7 lb), cock 3.17–3.62 kg (7–8 lb).

Origin: Between about 1850 and 1875, Cochin, Malay, Brahma and Croad Langshan fowls arrived in Holland from Asia and were crossed with local fowl. The ensuing Barnevelder breed originated in the Barneveld area of Holland slightly prior to 1914, with birds imported into Britain in 1921.

Barred Rock

This is the most popular variety of the Plymouth Rock breed, producing good eggs and ideal for the table. It is a heavy, soft-feather breed, handsome and with a plump body, yellow skin and narrow, clean barring that easily distinguishes it. The body is long, broad and deep. Young chicks are dark grey to black, with some white patches on the body and head. In addition to the Barred Rock, there are six other varieties of the Plymouth Rock breed – Blue, Buff, Columbian, Partridge, Silver-pencilled and White.

Temperament: Docile, quiet and well mannered – ideal as pets for children. Its prolific egg-laying abilities are not discouraged by cold winter weather.

Eggs: 220 each year, sometimes more.

Egg colour: Light to medium brown, with rich yellow yolks.

Weight: Hen 3.4 kg (7½ lb), cock 3.6 kg (8 lb).

Origin: It dates back to the 1800s in New England (USA) and originates from crosses between Dominiques and Black Javas; it became very popular in the 1920s and enthusiasm for it has not declined.

Black Australorp

Sometimes just known as Australorp, this heavy, soft-feather breed has glossy-black feathers with a deep, lustrous green sheen. The body is large and similar to the Orpington, but deeper and longer, and with flatter feathering. It is a dual-purpose breed, producing eggs as well as being ideal for the table. Birds are highly active, with dark, pronounced eyes. The face is unfeathered and with white skin, white soles to the feet and similarly coloured toe nails.

Temperament: Placid and friendly.

Eggs: 200 each year, often more (see below).

Egg colour: Tinted brown.

Weight: Hen 2.9–3.6 kg (6½–8 lb), cock 3.4–4.5 kg (7½–10 lb).

Origin: The breed was developed in Australia from Black Orpingtons and introduced into Britain in 1921. The name Australorp is claimed to be an abbreviation for Australian Black Orpington, although there is diverse opinion about this.

Australorp hens are famed for their egg-laying ability and, in 1922–23, six hens set a world record for laying eggs; each produced an average of 309 eggs over a period of 365 days.

Black Rock

This hybrid has dense, all-weather plumage which not only provides good insulation from cold and wet weather but makes the birds less susceptible to red mites. The plumage is predominantly black, with variable amounts of chestnut colouring around the neck.

Temperament: Friendly and docile, not easily stressed and therefore they do not usually require de-beaking. Their highly developed immune system makes them ideal for free-range activities, as well as ensuring a long, productive, egg-laying life.

Eggs: 280 each year – often more. The shell quality and colour remain constant throughout a hen's life.

Egg colour: Brown.

Weight: Hen 3.4 kg (7½ lb), cock 3.6 kg (8 lb)

Origin: First-cross hybrids originating from the Harco/Arbor Acres breeders of North America and created from selected strains of Rhode Island Red males and Barred Plymouth Rock females. They are now sold throughout the world, usually through licensed poultry breeders who have improved the breed while keeping its essential characteristics.

Black Silkie

In most of North America this light, soft-feather breed is bantam-sized, while in Europe it is larger and classified as a large fowl light breed. Silkies appear in two forms – Bearded and Non-bearded. Bearded types have an extra muff of feathers under the beak area that covers the earlobes. They have five toes on each foot, walnut-type combs, dark wattles and blue earlobes. In addition to Black Silkies, there are white, blue, gold and partridge varieties.

Temperament: Lively and friendly and ideal as pets for children. They are poor layers, but ideal as mothers for incubating their own eggs as well as those of other breeds. They are unable to fly so can be kept in areas with low fencing. They do little damage to gardens.

Eggs: 80–100 each year.

Egg colour: Tinted or cream.

Weight: Hen 1.4 kg (3 lb), cock 1.8 kg (4 lb).

Origin: An old breed; its earliest surviving account comes from the Venetian traveller Marco Polo (c.1254–1324) who recorded a furry chicken in Asia. Later, in 1599, Ulisse Aldrovandi, an Italian naturalist and writer, wrote of 'wool-bearing chickens', and ones 'resembling black cats'!

Bluebelle

This hybrid has pretty, blue-grey plumage, ranging from a pale grey to a dark slate-coloured blue.

Temperament: Placid and easy to handle, they are ideal as pets for children.

Eggs: 250 each year.

Egg colour: Pale brown.

Weight: Hen 3.4 kg (7½ lb), cock 3.6 kg (8 lb).

Origin: This distinctive hybrid originated in the Czech Republic and was derived from a Maran and a Rhode Island Red. These chickens are sometimes known as Meadowsweet Bluebells. It is an autosex hybrid and cannot breed true; its offspring will revert to type according to the cockerel used to fertilize them. It is therefore essential to buy fresh hens from a recognized supplier of the breed. Therefore, it is no good putting a cockerel in with your hens in the anticipation of producing Bluebelle chicks, as they will not breed true to type. This ruling applies to other hybrid chickens, including Black Rock.

Blue Laced Wyandotte

Heavy, soft-feather breed, with a deep chest and rounded rear end, creating an attractive, well-defined curved appearance. The ground colour of the feathers is red-brown, with the edges marked in blue.

Temperament: Alert, active, docile, graceful and ideal for showing as well as laying. Like all variations of Wyandotte, the Blue Laced Wyandotte is suitable as a pet for children. Its temperament suits it to both free-range conditions and in a run.

Eggs: 200–240 each year.

Egg colour: Light brown to brown – sometimes tinted.

Weight: Hen 2.7 kg (6 lb), cock 3.8 kg (8½ lb).

Origin: The Wyandotte breed originated in the USA in the form of the Silver-laced Wyandotte, where the breed was formally recognized in 1883. Later, the Blue Laced Wyandotte appeared, together with many other coloured forms, including the Gold Laced Wyandotte (page 17), the Silver Laced Wyandotte (page 21) and the Silver Pencilled Wyandotte (page 22). All of these handsome breeds have become widely popular.

Blue Silkie

In most of North America, this light, soft-feather breed is bantam-sized, while in Europe they are larger and classified as a large fowl light breed. Silkies appear in two distinct forms – Bearded and Non-bearded. Bearded types have an extra muff of feathers under the beak area that covers the earlobes. They have five toes on each foot, walnut-type combs, dark wattles and blue earlobes. Blue Silkies are light bluish in colour and, some people claim, have the look and feel of Persian cats. There are also White, Black, Gold and Partridge varieties.

Temperament: Lively and friendly and ideal as pets for children. They are poor layers, but ideal as mothers for incubating their own eggs as well as those of other breeds. Silkies are unable to fly so can be kept in areas with low fencing. They do little damage to gardens.

Eggs: 80–100 each year.

Egg colour: Tinted or cream.

Weight: Hen 1.4 kg (3 lb), cock 1.8 kg (4 lb).

Origin: An old breed, developed in the Orient and, most likely, in Japan.

Buff Cochin

A heavy, soft-feather, extra-large breed with plush, soft and mellow, buff-toned, broad and short feathers. The birds have a rounded and broad shape and are said to suffer from heart problems. Their legs are yellow, their earlobes red and their eyes a reddish-bay. The legs and feet are covered in plumage, which can be a problem where a bird is kept in long, wet grass.

Temperament: Friendly, docile and moderately active, but with a sedate, lazy nature which makes them good pets, especially as they live for 8–10 years. They do not fly and therefore require only a low fence. They are motherly and broody .

Eggs: 140 each year.

Egg colour: Brown – usually tinted.

Weight: Hen 3.8 kg (8½ lb), cock 4–4.5 kg (9–10 lb).

Origin: The Cochin came from China in the early 1850s, where it was known as the Cochin-China or Chinese Shanghai. Apart from the Buff Cochin, there are Black, Blue, Cuckoo, White, Partridge and Grouse-coloured forms.

Buff Orpington

Dual-purpose chicken; heavy, soft-feather breed producing eggs as well as being ideal for the table. It has a neat head and a single comb. There are several varieties of Orpingtons. The dark-coloured ones have dark eyes and legs, and the paler ones have red eyes and white legs. The amount of sun they receive influences their colouring, and therefore in hot areas with strong sunlight you need to provide the birds with light shade. They are rather heavy birds and usually unable to fly. They make excellent mothers.

Temperament: Docile but active and ideal as family pets as well as egg laying. They like to be handled and it is claimed that they will come when their name is called.

Eggs: 140–160 each year – sometimes more. They do not stop laying in winter.

Egg colour: Light brown.

Weight: Hen 3.6 kg (8 lb), cock 4.5 kg (10 lb).

Origin: An English breed, named after a village in Kent and introduced in 1894 by William Cook. He bred a black variety in 1886, and a white one in 1889.

Buff Sussex

Dual-purpose, heavy, soft-feather breed producing eggs as well as being ideal for the table. Apart from the Buff Sussex, there are several other varieties, including Brown, Light, Red, Speckled, Silver, White and Coronation. They have white legs and feet, and red earlobes. The eyes of dark-coloured varieties are red, while the lighter ones are orange. The legs are short, strong and free from feathers, and the back is broad and flat.

Temperament: Alert, docile and adaptable; ideal for confined places. They are also superb in free-range situations, as they naturally forage for food.

Eggs: 180–200 each year – sometimes more.

Egg colour: Cream to light brown – and tinted.

Weight: Hen 3.2 kg (7 lb), cock 4 kg (9 lb).

Origin: Derived from Old English fowls which were bred for their meat and eggs in Victorian times. The Sussex breed has made an important contribution to the modern poultry industry and its ancestry can be found in many breeds.

Cream Legbar

Apart from the Cream Legbar, there are Gold and Silver varieties of this light and slightly rare breed. They are firm, muscular birds, with a wedge-shaped body, broad at the shoulders and tapering towards the rear. They have strong beaks, with an erect, large comb composed of 5–7 even spikes. The crest, which lies towards the back of the head, is smaller in the male than in the female. The face is red and smooth, the beak yellow, and with pendant white or cream earlobes. Legs and feet are yellow, unfeathered and with four toes.

Each male chick has a pale dot on its head and with little or no eye barring, while each female chick has a dark brown or black stripe on its head, which continues down the body, and with clear eye barring. This enables unrequired male chicks to be readily identified and removed.

Temperament: Sprightly and alert.

Eggs: 170 each year.

Egg colour: Usually blue, but sometimes olive.

Weight: Hen 2–2.7 kg (4½–6 lb), cock 2.7–3.4 kg (6½–7 lb).

Origin: It is a cross between the Brown Leghorn and the Barred Plymouth Rock, with the addition of some Araucana blood.

Croad Langshan

Heavy, soft-feather breed with a large body and a deep and long breast. The back is rather long and sloping, with a tail that rises sharply from the back, giving a characteristic U-shape. The head appears to be small when compared to the body and the beak is light to dark horn.

Temperament: They are adaptable and do well in free-range situations as well as when confined. They are suited to warm climates but, in general, are not too fussy. The hens are good mothers and their ability to lay eggs is not diminished in winter.

Eggs: 180 each year – sometimes less.

Egg colour: Brown, with a plum-like bloom.

Weight: Hen 3.2 kg (7 lb), cock 4 kg (9 lb).

Origin: The Langshan has its origins in the Langshan District just north of the Yangtse-Kiang River in China. In 1872, Major F.T. Croad introduced them into Britain and the progeny of these fowls became known as Croad Langshan. They were imported into North America in 1878 and soon developed into several varieties – Black, White and Blue. In 1879 the breed was taken to Germany and these became known as the German Langshans.

Cuckoo Maran

Dual-purpose, heavy, soft-feather breed; hens have large legs and produce excellent meat for the table. Medium length bird, with a good width and depth for meat production. The head is neat, with a single comb, and the tail high; it has orange-red eyes, red earlobes and white legs. Apart from the Cuckoo Maran, there are several other varieties. These include the Copper, Black, Black Copper, Black-tailed Buff, White and Colombian forms.

Temperament: Active yet docile and graceful.

Eggs: 150–180 each year.

Egg colour: Deep chocolate-brown.

Weight: Hen 3.2 kg (7 lb), cock 3.6 kg (8 lb).

Origin: The breed originated in France and takes its name from the French town of Marans, northeast of La Rochelle in western France. In its native area, it thrives in rough or marshy terrain and therefore does well in free-range conditions.

Faverolles

Dual-purpose, heavy and soft-feather breed, ideal for eating and producing medium-sized eggs. It is especially good for laying eggs during the winter. It has a wide body with a broad, round head and reddish-bay eyes. Head adornments include a beard and side whiskers. The pinkish legs are sparsely feathered, with the feathering mainly on the outer of the five toes. The most common colour is Salmon, with varieties in Black, White, Blue, Buff, Cuckoo and Ermine.

Temperament: Quiet, friendly and gentle – ideal as pets for children. They are alert and active, the hens making very good mothers. The breed does well in free-range situations, as well as in an enclosure.

Eggs: 160 each year – often more.

Egg colour: Light brown or cream-tinted.

Weight: Hen 2.9–3.4 kg (6½–7½ lb), cock 3.6–4 kg (8–9 lb).

Origin: Originated in the village of Faverolles in northern France in the middle of the 19th century, with several breeds in their ancestry including Dorking, Brahma Crvecoeur, Houdan, Coucou de Rennes and, possibly, the Cochin.

French Copper Maran

Heavy, soft-feather breed with a dual-purpose role, producing large eggs and excellent meat for the table. Medium length bird, with a good width and depth for meat production. The head is neat, with a single comb, and the tail high. It has orange-red eyes, red earlobes and white feet. Day-old males have a large white spot on the top of the head (females a smaller one); this makes sexing chicks fairly easy if it is done early. Apart from the Copper Maran, there are several other recognizable varieties, such as Black, Black Copper, Black-tailed, White, Colombian, Wheaton Maran and Cuckoo Maran.

French Black
Copper Maran

Temperament: Active and graceful.

Eggs: 170–180 each year.

Egg colour: Very deep brown.

Weight: hen 3.2 kg (7 lb), cock 3.6 kg (8 lb).

Origin: The breed takes its name from the French town of Marans, northeast of La Rochelle in western France, and was imported into England in about 1929.

French Wheaton Maran

Heavy, soft-feather, dual-purpose breed, producing large eggs and excellent meat for the table. Medium length bird, with a good width and depth for meat production. The head is neat, with a single comb, and the tail high. It has orange-red eyes, red earlobes and white feet. Day-old males have a large white spot on the top of the head (females a smaller one); this makes sexing chicks fairly easy if it is done early. Apart from the French Wheaton Maran, there are several other recognizable varieties, such as Copper, Black, Black Copper, Black-tailed, White and Colombian.

Temperament: Active and graceful.

Eggs: 170–180 each year.

Egg colour: Deep brown.

Weight: Hen 3.2 kg (7 lb), cock 3.6 kg (8 lb).

Origin: The breed takes its name from the French town of Marans, northeast of La Rochelle in western France, and was imported into England in about 1929.

Frizzle

Distinctive and unusual heavy breed, with feathers that curl outwards rather than lying flat along the body, producing a pretty bird. They are primarily used for exhibition, but also lay eggs and can be kept for their meat. When hatched, the chicks have normal feathers, but soon the wing feathers develop and turn outwards. There are several different varieties, in colours including white, black, blue and buff. If you breed Frizzles, not all of their offspring will have frizzled feathers. However, if a non-frizzled bird is bred with a frizzled one, the progeny will have a good chance of being frizzled.

Temperament: Gentle and ideal for free-range conditions or in a caged run.

Eggs: 160 each year.

Egg colour: Creamy-white or brown-tinted.

Weight: Hen 2.7 kg (6 lb), cock 3.6 kg (8 lb).

Origin: It is thought to have originated in Southeast Asia, Java and the Philippines.

Gold Brahma

Heavy, soft-feather breed, producing large, tall and stately birds with broad backs and deep bodies; short tails and feathered legs. Also, a small head with beetle brows and a triple comb.

Temperament: Friendly, calm and ideal as pets for children, as well as for exhibition. They are not skittish or easily frightened and generally not aggressive to humans, especially if regularly handled from when they are young.

Eggs: 80–100 each year – sometimes slightly more.

Egg colour: Brown and speckled.

Weight: Hen 4 kg (9 lb), cock 4.5 kg (10 lb).

Origin: Brahmas are an Asiatic breed, originating in the Brahmaputra region of India and known as Gray Chittagongs. They were taken to North America in 1846 and initially known as Brahma-Pootra, which was then shortened to Brahma. The breed was developed in North America and introduced into England in 1852, when nine birds were presented to Queen Victoria. For many decades there were only two varieties, the Light and the Dark. Later, others appeared, including the White and Buff Columbian. In all of these, the eyes, comb and earlobes are red, with the legs a distinctive bright yellow.

Gold Laced Wyandotte

With its deep chest and rounded rear end, this heavy, soft-feather breed has an attractive, well-defined and curved appearance. It is a beautiful gold-coloured bird, with a black edge to its feathers that creates a black, scalloped look all over its body.

Temperament: Alert, active, docile, graceful and ideal for showing as well as laying. Like all variations of Wyandotte, the Gold Laced Wyandotte is an ideal pet for children. Its temperament suits it to free-range situations as well as in a run.

Eggs: 200–240 each year.

Egg colour: Light brown to brown – sometimes tinted.

Weight: Hen 2.7 kg (6 lb), cock 3.8 kg (8½ lb).

Origin: The Wyandotte breed originated in the USA in the form of the Silver Laced Wyandotte, where the breed was formally recognized in 1883. Later, the Gold Laced Wyandotte appeared, together with the many other coloured forms, including the Blue Laced Wyandotte (page 13), the Silver Laced Wyandotte (page 21) and the Silver Pencilled Wyandotte (page 22). All of these handsome breeds have become widely popular.

Hamburg

A light, soft-feather, distinctive breed that always looks attractive. There are several superb varieties, including the White, Black, Silver Pencilled, Golden Spangled and Golden Pencilled. However, it is most often seen in the Silver Spangled form, with a white ground colour and each feather finished in a black spangle.

Golden Pencilled Hamburg

Temperament: Alert and active, with a friendly but flighty disposition. It dislikes confinement and is better as a free-ranger than in a run.

Eggs: 240 each year.

Egg colour: White, with a glossy shell.

Weight: Hen 1.8 kg (4 lb), cock 2.3 kg (5 lb).

Origin: An old, light breed that originated in Holland and was known in Britain more than 300 years ago, where the birds were referred to as Moonies, Dutch Everyday Layers and Everlayers. Much of the breed's present development is owed to British chicken breeders.

Leghorn

A light, soft-feather breed, with white earlobes, yellow legs and red eyes. Their plumage is soft and silky and the legs long and featherless, with four toes on each foot. The hens are good layers and do not go broody. They are usually noisy birds, active and well able to fly over low fences – ensure the enclosure is at least 1.8 m (6 ft) high. Leghorn varieties include White, Red, Black-tailed Red, Light Brown, Dark Brown, Black, Cuckoo and Partridge.

Temperament: They are equally suited to roam freely or in a large run. This breed does best in warm climates, as the combs on male birds are likely to be damaged in severe weather (a coating of petroleum jelly provides protection).

Eggs: About 200 each year – often more.

Egg colour: White.

Weight: Hen 2.5 kg (5½ lb), cock 3.4 kg (7½ lb).

Origin: Leghorns originate from the Port of Leghorn in Italy and were taken to North America in 1853, where initially they became known as Italians. They were introduced into Britain in the mid-1800s.

Light Brahma

Heavy, soft-feather breed. Large, tall and stately, with broad back, deep body, short tail, feathered legs, small head, beetle brows and triple comb. The plumage is profuse, giving the appearance of a larger bird. The base colour is white, with black-and-white hackle feathers. The main tail feathers are black, with white side hangers. For many decades there were only two varieties: Light and Dark. Later, others appeared, including the Gold Brahma, White, Dark and Buff Columbian. In all of these, the eyes, comb and earlobes are red and the legs bright yellow.

Temperament: Friendly, calm and ideal as pets for children, as well as for exhibition. They are not skittish or easily frightened and generally are not aggressive to humans, especially if regularly handled from when young.

Eggs: 80–100 each year – sometimes slightly more.

Egg colour: Brown and speckled.

Weight: Hen 4.08 kg (9 lb), cock 4.5 kg (10 lb).

Origin: Brahmas are an Asiatic breed, originating in the Brahmaputra region of India and known as Gray Chittagongs.

Light Sussex

Heavy, soft-feather breed, with the shoulder feathers displaying a distinctive black stripe down the centre; the shaft is also black. The wings are white, with black flight feathers, and the remaining body feathers are pure white. Feet, toes and toenails are white. The whole bird is graceful, with a long, broad, flat back and tail feathers at an angle of 45 degrees. The head is neat, with a medium-sized, single comb. Apart from the Light Sussex, the Speckled and Red varieties are recognized in North America, while in England other varieties include the Buff, Brown and Silver.

Temperament: Alert but docile and adaptable to confined places as well as suitable for free-range situations.

Eggs: 220–220 each year – sometimes more.

Egg colour: Cream to light brown.

Weight: Hen 3.2 kg (7 lb), cock 4 kg (9 lb).

Origin: The original Sussex chickens are claimed to have been in existence for 2,000 or more years and were mainly bred for their meat. It was a hardy breed and could survive in all weathers. It is thought to derive from the Ardennes region of Belgium and was introduced into Britain by the Romans. The oldest variety of the Sussex breed is the Speckled; later, Brahma, Cochin and Silver Grey Dorking breeds were used to produce the Light Sussex.

Minorca

Originally known as Red-faced Spanish, this light breed is the largest and heaviest of the Mediterranean breeds. It is very distinctive, with a bright red comb and large, almond-shaped, white earlobes. It has a strong body and an upright stance. Black is the main colour, but there are two other varieties – White and Blue.

Temperament: It is best kept as a free-ranger, but is just as happy in a large run. Moderate to warm climates suits it best. It is claimed to have been the favourite breed of chickens kept by the legendary 'Lawrence of Arabia', T.E. Lawrence.

Eggs: 240 each year.

Egg colour: White.

Weight: Hen 3.2 kg (7 lb), cock 3.6 kg (8 lb).

Origin: It gains its name from Minorca, from where the birds were first exported. It can be traced back to 1780, or earlier, but did not become popular until the early 1900s.

New Hampshire Red

Heavy, soft-feather, dual-role chicken, producing slightly fewer eggs than its ancestry line, the Rhode Island Red, but more meat. Its plumage is fluffy and full and more orange than the Rhode Island Red; the male has light yellow-orange hackle feathers. Their legs are yellow and the lower thighs large and muscular. They are not high fliers and therefore do not need high fencing.

Temperament: Placid and friendly and easy to tame; they are at home in a run or as free-rangers. Hardy in nature and well able to survive low temperatures while still laying plenty of eggs.

Eggs: 200–220 each year – although sometimes fewer.

Egg colour: Brown and tinted.

Weight: Hen 2.9 kg (6½ lb), cock 3.8 kg (8½ lb).

Origin: Derived from Rhode Island Reds and altered through selective breeding into a distinctive breed in the early 1900s; it was accepted as a new breed in 1935.

Partridge Silkie

In most of North America, this light, soft-feather breed is bantam-sized, while in Europe it is larger and classified as a large fowl light breed. Silkies appear in two distinct forms – Bearded and Non-bearded. Bearded types have an extra muff of feathers under the beak area that covers the earlobes. They have five toes on each foot, walnut-type combs, dark wattles and blue earlobes. This pretty bird has plumage that displays a medley of partridge-type colours. In addition to Partridge Silkies, there are White (page 22), Blue (page 13) and Gold varieties.

Temperament: Lively, friendly and ideal as pets for children. They are poor layers, but ideal as mothers for incubating their own eggs as well as those of other breeds. They cannot fly, so can be kept in areas with low fencing. They do little damage to gardens.

Eggs: 80–100 each year.

Egg colour: Tinted or cream.

Weight: Hen 1.4 kg (3 lb), cock 1.8 kg (4 lb).

Origin: An old breed; its earliest surviving account comes from the Venetian traveller Marco Polo (c.1254 to 1324) who recorded a furry chicken in Asia.

Plymouth Rock

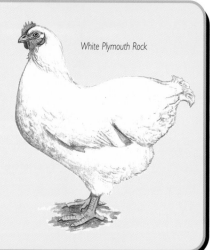

White Plymouth Rock

Popular breed, originating in North America, with the Barred Rock (see page 11) being the first variety. It is a dual-purpose chicken, ideal for laying as well as for eating. In addition to the Barred Rock, there are other popular varieties, including White and Buff. They are handsome, with long, broad, deep bodies.

Temperament: Docile, friendly and easily handled; ideal as pets for children. The birds are suitable for free-range activities, as well as in runs. Its egg-laying ability is not diminished by cold winter weather.

Eggs: 200 each year.

Egg colour: Light to medium-brown, sometimes with a touch of pink.

Weight: Hen 3.4–3.6 kg (7½–8 lb), cock 4.5 kg (10 lb).

Origin: Several breeds are in its ancestry, including the Cochin, Brahma, Java, Dominique and Minorca.

Redcap

Dual-purpose, light breed, ideal for both meat and egg production. It has a broad body and solidly fleshed breast. The close-fitting wings are red-brown, with beetle-green webbing. The legs and feet are lead-coloured. However, the very large, red rose-comb is its most striking feature, with a straight 'leader' pointing backwards, and fine spikes.

Temperament: Hardy and active, with a somewhat flighty nature, the Redcap is at its happiest when free-ranging and foraging. It is a good flier and requires a high fence. Its comb is sometimes adversely affected by cold weather.

Eggs: 180–240 each year.

Egg colour: White.

Weight: Hen 2.5 kg (5½ lb), cock 2.9 kg (6½ lb).

Origin: Native to Britain and often known as the Derbyshire Redcap. Breeds thought to be in its ancestry include Dorkings, Old English Pheasant Fowl, Golden Spangled Hamburg and Black-breasted Red.

Red Dorking

Dual-purpose, heavy, soft-feather breed, ideal for producing eggs as well as meat for the table. Indeed, Dorkings were originally highly prized for their white meat. The red variety of the Dorking is very dark and a distinctive bird.

Temperament: Usually calm, quiet, shy and docile.

Eggs: 120–130 each year – sometimes more. They tend to lay more eggs in spring and summer than during autumn and winter.

Egg colour: White – sometimes with a pinkish tinge.

Weight: Hen 3.2–4.5 kg (7–10 lb), cock 4–5.4 kg (9–12 lb).

Origin: The original Dorking is claimed to have been introduced from Italy to the British Isles by the Romans and said to be the first epicurean bird in England; capons were raised for banquets. It was mentioned during the reign of Julius Caesar (100–44 BCE) by the Roman writer Columella, who wrote about farm animals.

Rhode Island Red

Dual-purpose, heavy, soft-feather breed, mainly used to produce eggs. Medium to heavy build, with dark, rust-coloured feathers, although maroon to nearly black types are known. They have red-orange eyes, red-brown beaks and yellow feet.

Temperament: Hardy, friendly, alert and good-natured. They can become aggressive if annoyed, although they generally make good pets for children.

Eggs: 200–220 each year – sometimes more. They continue to lay eggs even when the temperature drops, although when it is below 0°C (32°F) laying is reduced.

Egg colour: Light to dark brown.

Weight: Hen 2.9 kg (6½ lb), cock 3.4–3.8 kg (7½–8½ lb).

Origin: North American breed admitted to the American Standard of Perfection in 1904. First exhibited in 1880 in South Massachusetts and developed from the Asiatic black-red fowls of Shanghai, Malay and Java. It has proved to be an excellent layer and is used by many breeders to improve their stock. Indeed, many modern hybrid hens have Rhode Island Red fathers; the characteristic of prolific laying is passed down through the male line.

Silver-grey Dorking

Heavy, soft-feather chicken with short, sturdy legs, rectangular body and five toes (most chickens have four). In early years it was mainly kept for its white meat; nowadays, it is noted for both eggs and meat. Cockerels have upright combs, while the comb on a hen is large, floppy and falls to one side. Because of the comb's relatively large nature, this breed may need extra protection in cold weather. Related variations include Red Dorking (see page 20) and White Dorking.

Temperament: Usually calm, quiet, stately, shy and docile; a pretty hen that endears itself to chicken enthusiasts and is ideal for showing.

Eggs: 120-130 each year – sometimes more.

Egg colour: White, with a pinkish tinge.

Weight: Hen 3.2–4.5 kg (7–10 lb), cock 4–5.4 kg (9–12 lb).

Origin: The original Dorking is claimed to have been introduced from Italy to the British Isles by the Romans and is said to be the first epicurean bird in England; capons were raised for banquets.

Silver Laced Wyandotte

Heavy, soft-feather breed, with a deep chest and rounded rear end; it has a well-defined and curved appearance. The feathers are silvery-white, with black edges, an effect known as lacing. The tail is black, and the feet yellow. Birds mature rapidly and the meat is good. Indeed, birds are kept for the table as well as for their eggs.

Temperament: Alert, active, docile, graceful and ideal for showing as well as laying.

Eggs: 200–240 each year.

Egg colour: Light brown to brown – sometimes tinted.

Weight: Hen 2.7 kg (6 lb), cock 3.8 kg (8½ lb).

Origin: This breed was the first of the Wyandotte range of chickens. It originated in North America and was standardized in 1883. There are eight different variations of Wyandottes recognized by the American Poultry Association; four of them are featured in this A–Z of chicken breeds.

Silver Pencilled Wyandotte

Heavy, soft-feather breed. With its deep chest and rounded rear end, this breed has an attractive, well-defined and curved appearance. It partly resembles a partridge; the hen has a silver undercolour, while the male is white but with bits of black that graduate to totally black in the tail and wings. Birds mature rapidly and the meat is good. Indeed, birds are kept for the table as well as their eggs.

Temperament: Alert, active, docile, graceful and ideal for showing as well as the production of eggs. Like all variations of Wyandotte, the Silver Pencilled Wyandotte is an ideal pet for children.

Eggs: 200–240 each year.

Egg colour: Light brown to brown – sometimes tinted.

Weight: Hen 2.7 kg (6 lb), cock 3.8 kg (8½ lb).

Origin: See Silver Laced Wyandotte (page 21).

Welsummer

Light, soft-feather, large and upright breed, with a deep and full breast, large tail and a broad back. The beak is strong and short; almond-shaped earlobes and a single comb. The eyes are reddish-bay, while the yellow legs fade to pale yellow in summer. They are happy when in an enclosed run, or as free-rangers.

Temperament: Active, alert, friendly and easily handled. They do become broody, but not usually until late spring; unfortunately, they are not good mothers.

Eggs: About 200–240 each year; fewer eggs in winter than summer.

Egg colour: The brown, mottled eggs are some of the largest of chicken eggs. They are sometimes described as 'a rich, deep, flower-pot red'; however, they are certainly more dark brown than red.

Weight: Hen 2.72 kg (6 lb), cock 3.2 kg (7 lb).

Origin: Developed from 1900 to 1913 in the area alongside the river Ysel, to the north of Deventer, Holland, and named after the village of Welsum. Its initial development owes much to breeds such as the Partridge Cochin, Partridge Wyandotte and Partridge Leghorn. It was taken to Britain in 1928, primarily for its large, brown, mottled eggs.

White Silkie

In most of North America, this light, soft-feather breed is bantam-sized, while in Europe it is larger and classified as a large fowl light breed. In addition to the white variety, there are Black, Blue, Buff, Grey and Partridge forms. They all have walnut-type combs, dark wattles, blue earlobes, and five toes on each foot.

Temperament: Calm, docile, friendly disposition and ideal as pets for children. Although poor layers, they are ideal as mothers. They are unable to fly so can be kept in areas with low fencing. They do little damage to gardens.

Eggs: 80–100 each year.

Egg colour: Tinted or cream.

Weight: Hen 1.36 kg (3 lb), cock 1.81 kg (4 lb).

Origin: Sometimes misspelled as Silky, its origin is uncertain but the breed has been known for several hundred years. They arrived in Europe about 200 years ago and are said to have originated in India, China or Japan. Amusingly, at that time they were claimed to be crosses between chickens and rabbits!

BANTAM BREEDS

There is a choice between large chickens (sometimes known as large fowl) and bantams (see page 10). A range of popular 'true' bantams is featured on pages 24–27.

However, there are many large, normal-sized breeds that have been bred to produce bantam-sized versions, and these are indicated below.

LARGE-FOWL BREEDS WITH BANTAM-SIZED VERSIONS

- Araucana
- Australorp
- Barnevelder
- Brahma
- Cochin
- Croad Langshan
- Dorking
- Faverolles
- Frizzle
- Hamburg
- Leghorn
- Maran
- Minorca
- New Hampshire Red
- Orpington
- Plymouth Rock
- Rhode Island Red
- Silkie
- Sussex
- Welsummer
- Wyandotte

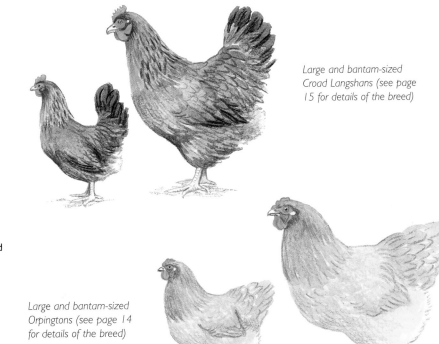

Large and bantam-sized Croad Langshans (see page 15 for details of the breed)

Large and bantam-sized Orpingtons (see page 14 for details of the breed)

Large and bantam-sized Welsummers (see page 22 for details of the breed)

Confusing times

There are some breeds which are known in one country as large fowl, while in another as bantam. This is often because as different countries have bred and developed breeds of chickens, their sizes have become altered to accommodate specially selected characteristics.

PEKIN BANTAMS

The Pekin is a 'true' bantam, which means it does not have a large-fowl equivalent. It is only 20–30 cm (8–12 in) high, even with its head upright. Some bantam enthusiasts believe the Pekin to be a miniature Cochin, but there are some differences in its stance.

The Pekin has a characteristic 'tilt', with a sloping and forward carriage, and the breast of the female almost touches the ground. Also, the body is round and has the appearance of a 'walking tea cosy'. It has a single comb and feathered legs (especially in the cockerel) where the feathers tend to spread out from the sides.

There are many varieties of Pekin Bantams, in a wide colour range (see below and page 25).

Temperament: Docile, gentle and friendly, making good pets for the family and children. They are good layers of small eggs, but tend towards broodiness and therefore make good mothers.
Eggs: Small; breed mainly kept for its ornamental qualities.
Egg colour: White to tinted.
Weight: Hen 570–680 g (20–24 oz), cock 680–790 g (24–28 oz).
Origin: There are many stories about this breed's ancestry, with claims it was stolen from a private collection of the Emperor of China in about 1860. Another suggestion is that they were presented to Queen Victoria during the 1830s. These birds were then crossed with other breeds to produce the now popular Pekin Bantams.

Columbian Pekin Bantam

Cuckoo Pekin Bantam

Lavender Pekin Bantam

Pekin Bantam clubs

Enthusiasm for Pekin Bantams is worldwide and increasing each year. This has encouraged the initiation of many Pekin clubs and societies – some local, others national – for the many dedicated enthusiasts of this breed. These associations encompass countries throughout Europe as well as North America and Australia. A glance on the internet will soon provide you with the addresses of Pekin Bantam clubs in your area or country.

Because of the increasing popularity of this breed, you should not have any problem in locating a source of birds – but always go to a reputable supplier. Remember that each year breeders of Pekin Bantams produce more colour variations; some may yet have to be properly recognized but you will nevertheless find those varieties listed on page 25.

Pekin Bantam colour range

The range of colour variations is amazingly wide, which indicates the large interest shown in this bantam, especially in North America. Below are the main colour variations you will encounter.

- Black
- Black Mottled
- Blue
- Buff
- Colombian
- Cuckoo

- Gold Partridge
- Lavender
- Partridge
- Red
- Silver Partridge
- White

Cochin Bantam colour range

The range of colours is wide, with further variations within each colour. Here are the basic colours:

- Barred
- Birchen
- Black
- Blue
- Buff
- Golden Laced

- Partridge
- Red
- Silver Pencilled
- Splash
- White

COCHIN BANTAMS

These resemble Pekin Bantams, but they do not have such a round and 'tea-cosy-like' appearance. Nor do they have a forward tilt, especially in the females.

Cochin Bantams have a dedicated following of enthusiasts and deserve more recognition. Their colour range is wide and often brighter in their variations than with Pekin Bantams. Few bantam enthusiasts would not be attracted by the bright plumage of Partridge and Birchen varieties, and the Golden Laced types look majestic in their rich and varied colourings. For more subdued colours, however, the Buff and White forms have few rivals.

Cochins were originally bred in China and taken to America and Britain in the mid-19th century. At that time they were large fowl, but later bantam versions were developed. They are pretty and become very tame. They are good mothers and able to cover a large number of eggs. When broody, they will accept the eggs of other breeds and become surrogate mothers very easily.

Confusion between Pekin Bantams and Cochin Bantams arises because in many countries Cochin Bantams are sold and known as Pekin Bantams.

Similarly to Pekin Bantams, there are many poultry clubs that specialize in Cochin Bantams, displaying them at shows and making them better known.

Golden-blue Cochin Bantam

White Cochin Bantam

Belgian

These are more complex than most bantam breeds, as there are three different forms: the d'Uccle (single comb, bearded, whiskers and feathered legs); the d'Anvers (bearded, rose comb, whiskers and clean legs); and the d'Watermael (similar to the d'Anvers but with three leaders from the back of its comb). In addition to these three varieties, they are seen in many colour variations, as well as a wide range of weights.

Temperament: Lively, jaunty and amusing bantams, with a low stance that results in a rather 'dumpy' but attractive carriage and appearance.

Eggs: Tiny; breed mainly kept for its ornamental qualities.

Egg colour: Creamy-white.

Weight: Hen 620 g (22 oz), cock 740 g (26 oz).

Origin: During the 1880s, breeders in the town of Uccle in Belgium started creating the d'Uccle breed by crossing the Bearded d'Anvers with the Booted Bantam. The Bearded d'Anvers has a longer ancestry, dating back to the 17th century.

Dutch

Dutch bantams are sometimes known as De Hollandse Krielan and originally Old Dutch. They are one of the smallest true bantams, with an upright stance, high and full breast and short back. The large and long wings are carried close to the body. It has a single red comb with five serrations, white earlobes, a short beak and slate-blue legs. There are several varieties, including Gold Partridge, Silver Partridge, Yellow Partridge, Blue-silver Partridge, Blue-yellow Partridge, Blue Partridge, Red-shouldered White, Cuckoo Partridge, Cuckoo, Black, White, Blue and Lavender. The American Bantam Association acknowledges more of these varieties than most other countries.

Temperament: Lively and jaunty birds, with a smart nature. They are good layers, but egg production is often limited to the summer months. These attractive hens make ideal mothers, although unable to cover a large number of eggs at one time.

Eggs: Tiny; breed mainly kept for its ornamental qualities.

Egg colour: Light tint.

Weight: Hen 400–450 g (14–16 oz), cockerel 480–540 g (17–19 oz).

Origin: They originated in Indonesia and were taken to the Netherlands in the 17th century by seamen in the East India Company.

Japanese

The Japanese Bantam, also known as Chabo, has a distinctive shape. It has a short back, high and upright tail, and very short legs which give it a wading gait. It has a large, evenly serrated, single comb. The body appears to tilt forward. There are several varieties, with the most popular being the Black-tailed White, Black and White, and Birchen. You might also see Black Breasted Red, Blue, Gray and Black-tailed Buff.

Temperament: It has a friendly disposition, allowing itself to be petted, lifted up and carried. However, male birds can be aggressive. Because they do not have a deep-pecking and grass-damaging tendency, in earlier times they were allowed to wander on lawns with short grass.

Eggs: Very small; breed mainly kept for its ornamental qualities.

Egg colour: Very small and creamy-white.

Weight: Hen 620 g (22 oz), cock 740 g (26 oz).

Origin: Developed in Japan and thought to have originated earlier in Southeast Asia, enjoying popularity in Malaysia, Java and Indonesia. The breed appeared in Japanese art in 1635, as well as in Dutch art of the same period. The name Chabo originated in Java as *Chabol*, meaning dwarf.

Nankin

Hardy bantam, slow to mature, with a high, arched neck and tail that is carried upright with feathers well spread out. The beak is strong, medium in length and attractively curved. The Nankin is seen in only one colour, buff, but is available in two comb forms – single or rose.

Temperament: They are not always friendly, especially the cocks. However, it is a breed that tolerates being kept in enclosures.

Eggs: Tiny eggs; breed mainly kept for its ornamental qualities.

Egg colour: Creamy-white.

Weight: Hen 620 g (22 oz), cock 680 g (24 oz).

Origin: One of the early true bantam breeds and thought to be used in the development of the Sebright bantam. It is of Asian origin, most likely from India. It is a breed that has been used to produce buff colours in other breeds.

Rosecomb

Sometimes known as Rose Comb Bantam, it slightly resembles the shape of the Dutch Bantam. It has a rose comb, square in front and ending in a long, pointed, slightly upward spike at the rear. Both the Black and the Blue varieties have slate-coloured shanks and toes; the White variety has pinkish-white ones. The breed has distinctive large, round, white earlobes. There are several varieties – Black, Blue (not laced) and White. Breeders are also developing other colours.

Temperament: They are surprisingly hardy in cool climates, as well as tolerating warm seasons. They are good fliers and tend to be very friendly as pets. However, the cocks can be aggressive.

Eggs: Tiny eggs; breed mainly kept for its ornamental qualities.

Egg colour: White to cream.

Weight: Hen 510 g (18 oz), cock 620 g (22 oz).

Origin: An old breed, thought to have originated in Britain and derived from Black Hamburgs and other bantams.

Sebright

Perhaps the most unusually coloured and distinctive of all bantam breeds. By 1899, two varieties had been established: the Gold and the Silver. The Gold variety is often described as rich and deep, while the Silver type appears more white than silver. The feathers of both these varieties are edged in black.

Temperament: Alert, long-lived and active, with a hardy nature and a liking for roosting in trees. They do not lay many eggs. They are difficult birds to raise and therefore not suitable for beginners to poultry keeping.

Eggs: Small; breed mainly kept for its ornamental qualities.

Egg colour: White or creamy-white.

Weight: Hen 570 g (20 oz), cock 620 g (22 oz).

Origin: A British breed, developed in the early 1800s by Sir John Sanders Sebright. The Sebright derives mainly from a cross between a bantam and a Poland (sometime known as a Polish), which is said to have originated in an area between Sweden and the Baltic States and then under Polish rule.

Ways to keep chickens

What are the choices?

There are several ways in which you can keep chickens, from methods that are solely commercial to others ideal for a small back yard and where you only want to keep a few chickens. A major influence on the way they are kept is the amount of land available. Free-ranging chickens have every possible opportunity to stretch their legs in idyllic conditions, but a chicken house and run in a back garden or yard can be just as productive and equally interesting.

FREE-RANGE SYSTEM

Essentially, a large amount of land is needed, where chickens can live in a field or, preferably, on land which has been reduced to stubble after being harvested; the chickens pick up seeds and rid the land of insect pests. This is idyllic but not within the possibility of many home poultry enthusiasts. However, it has a large number of star qualities, since for much of the year the birds are able to scavenge a large part of their food. In addition, the hens, through their own droppings (see pages 30–31) help maintain a field's fertility.

A hen house, often on wheels (see page 32) so that the benefits (manure and picking up insects) of chickens are spread over all of the entire area, is essential; the hen house will also keep your chickens safe at night.

As an alternative to a wheeled hen house, an ark (see page 32) is also suitable and this can be easily moved, usually by two people.

On the negative side of free-ranging is the risk of foxes attacking the chickens. Electric fences are ideal for keeping hens in and foxes out. (See page 33 for how to erect a non-electric fence.) Any eggs laid outside the ark and in hidden nests are also at risk of being taken by predators.

Where only a few hens are kept in small and wheeled hen houses or arks, they are especially at risk from low winter temperatures. Where there is a group of hens, the collective body warmth helps to raise the temperature. Low temperatures may decrease their ability to lay eggs. Conversely, high summer temperatures in open fields will harm them, so you must make sure that shade is available.

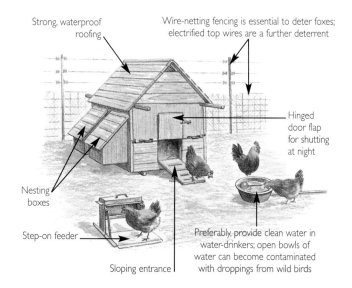

Strong, waterproof roofing

Wire-netting fencing is essential to deter foxes; electrified top wires are a further deterrent

Hinged door flap for shutting at night

Nesting boxes

Step-on feeder

Preferably, provide clean water in water-drinkers; open bowls of water can become contaminated with droppings from wild birds

Sloping entrance

Chickens are inquisitive and will drink from many sources. However, water is always best provided by clean water-drinkers.

FOLD METHOD

This is where chickens are kept in a poultry house with both its own and attached wire-netting enclosures. This creates a safe home, with the capability to be moved when the grass under it becomes excessively grazed, which can be every few weeks. Regularly moving the fold needs muscle power and time, but it does prevent land becoming excessively grazed. It also reduces the possibility of a build-up of parasites and diseases.

A variation on this fold system is a poultry house with scratching run (see page 32), but this is static.

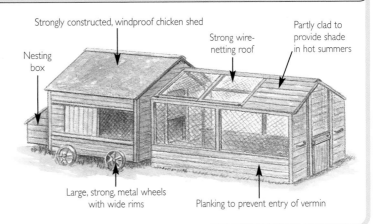

Strongly constructed, windproof chicken shed

Strong wire-netting roof

Partly clad to provide shade in hot summers

Nesting box

Large, strong, metal wheels with wide rims

Planking to prevent entry of vermin

SEMI-INTENSIVE SYSTEM

Unless you have a farm with a sound, unused building and a couple plots of land for chickens to graze on, this system will not be suitable for you. If you do have these facilities, however, it gives the chickens freedom to wander during the day and to be corralled inside at night on a straw-covered floor. In cold, wet weather the hens can be kept indoors, but throughout summer allowed to graze alternatively on the two plots of grass land. Alternating the areas of land helps to prevent one area becoming worn out or contaminated by pests and diseases.

DEEP-LITTER SYSTEM

Hens are homed in a large shed, without being allowed out of doors at all. Indeed, it is a variation on the semi-intensive system, with chickens kept on straw or litter that is topped up regularly and replaced annually. The system allows hens to walk freely although being confined indoors, but it does provide safety and warmth for them throughout the year.

Keeping hens constricted indoors does not appeal to many poultry enthusiasts and therefore a variation on the deep-litter system is to allow the hens to wander outdoors during daytime in a wire-netting enclosure with a thin covering of straw on the ground. This is method is sometimes known as the straw-yard system.

BATTERY SYSTEM

This is a solely commercial way of keeping chickens and not suitable for poultry enthusiasts who want to provide their chickens with a good life.

Housing your chickens

How much space do chickens need?

- For sleeping, allow a minimum of 30,000 cm^3 (1 cubic ft) per bird.

- For covered run and occasional garden access, each bird needs 1 m^2 (3 sq ft).

- Large chickens need about one-third more space than an average-sized chicken.

- Bantams need about one-third less space than an average-sized chicken.

- Allow space for six chickens – this will provide your family with eggs throughout the year. Three chickens will suit two people.

- Squeezing too many chickens into a hen house will cause them to squabble and peck at each other.

Choosing the right site

Getting the position of a hen house right will help to prevent problems that could arise later. Here are the main considerations:

- Good soil drainage – reduces risk of diseases and pests.

- Excellent air movement around the poultry shed.

- Convenient supply of clean, piped water.

- A site where summer smells will not be blown onto neighbouring land.

- Avoid positions at the bottom of a slope, where air becomes stagnant and bad drainage is likely. The air at the base of a slope is also much colder in winter than that found further up the slope.

- If the house is positioned on the side of a slope, ensure water running down the slope cannot run into the enclosure.

- The house should face the sun to take advantage of the sunlight – the pen will be warmer and drier.

- The closed or rear side of a chicken house should provide protection from prevailing cold winds.

- Where the main winter weather is coming from the west, the chicken house should face east. Conversely, where the main winter weather is coming from the east, face the chicken house towards the west.

- Essentially, the site and housing should provide chickens with dry feet, dry litter, fresh air and no draughts.

Chicken manure – its uses

It is useful in a garden?

Chicken manure is ideal for adding fertility to soil, especially as it is high in nitrogen, which is ideal for encouraging leafy growth in vegetable gardens. However, it does not have the soil-improving nature of manure from horses and pigs, which is usually accompanied by masses of decomposing straw. For that reason, chicken manure is best first added to a compost heap before being dug into the soil about a year later to improve its fertility and structure.

Chicken manure, when decomposed with vegetable garden waste, dramatically improves the soil and produces better crops.

POSSIBLE QUANTITIES

The amount of manure produced each day by a single chicken is often more than 112 g (4 oz). This means that in one year five chickens produce about 206 kg (456 lb) of manure. This is a welcome by-product of keeping chickens but plans must be ready for its use to avoid it creating a space and hygiene problem. It rapidly loses its benefits as a garden fertilizer and soil improver if left in a heap that is continually saturated with rain.

WHAT IS ITS VALUE IN A GARDEN?

All poultry manures are far richer in sulphate of ammonia, superphosphate and potash than farmyard manure such as that produced by pigs, horses and cattle. For example,

poultry manure contains more than twice as much nitrogen as that of pigs, and nearly four times more than cattle. The phosphate value is more than twice as much as pig manure, and more than five times that of cattle manure. As for potash, poultry manure contains more than twice that of pig manure and slightly more than cattle manure.

Part of this value as a fertilizer derives from the large amount of plant foods that quickly passes through poultry and is rapidly excreted.

Unlike mammals, poultry do not urinate and defecate separately, and therefore the semi-solid urine (white part of the droppings) can be easily seen. If this – when still fresh – comes into direct contact with plants it usually causes damage to stems and leaves by 'burning' them.

HOW CAN I USE CHICKEN MANURE IN A GARDEN?

Because its fertilizer value is rapidly reduced if exposed to weather – and especially to rain – it is best immediately added to a compost heap, where it acts as an activator in the decay of other materials in the heap. Spread it in 2.5–5 cm (1–2 in) thick coverings between layers of topsoil (which contains more bacteria and other soil organisms than subsoil) and soft garden waste such as grass cuttings and vegetable waste.

After a year in a compost heap, the material can be dug into the soil during autumn or winter digging. Alternatively, it can be used as a mulch – but ensure it does not come into contact with the stems of plants, as it may damage them.

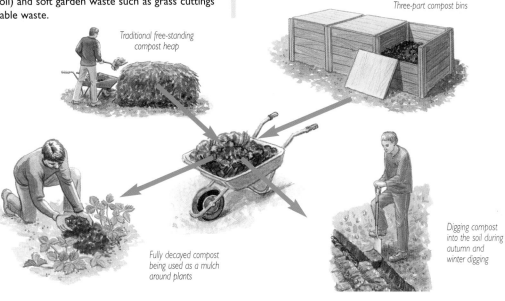

Three-part compost bins

Traditional free-standing compost heap

Fully decayed compost being used as a mulch around plants

Digging compost into the soil during autumn and winter digging

Can I mix lime with poultry manure?

No, they must be applied separately. If your soil is acidic and requires an application of lime to reduce its acidity, dig in poultry manure (as well as other animal manures) during autumn and early winter digging; then wait 2–3 months before dusting the soil with lime.

Warning

Wear a mask and goggles when cleaning out old chicken houses and runs. This is because dry and aged poultry manure can hold incubating spores of a human respiratory disease.

Also wear gloves and overalls to protect your hands and arms, as the manure is caustic and can burn skin. If your hands become contaminated, wash them immediately in hot, soapy water to which a disinfectant has been added.

Vegetables that especially benefit from a high-nitrogen diet

Vegetables that like well-decayed compost or manure dug into the soil when the land is being prepared for them during autumn and winter digging include:

- Beans (Runner and French)
- Capsicums
- Celery
- Cucumbers
- Endives
- Leeks
- Lettuces
- Marrows
- Peas
- Onions
- Spinach
- Sweetcorn
- Tomatoes

Poultry sheds and equipment

Several types of shed are available and their cost is influenced by the number of the chickens you intend to keep. Never constrict chickens – you should allow 0.2–0.3 m² (2–3 sq ft) of run space for each bird. Mobile housing ranges from arks that can be lifted by two people to large houses with wheels. Static housing ranges from ones where chickens can roam outside and within a fenced area to larger ones with a cage area attached.

TYPES OF POULTRY HOUSE

Ark

Ideal when a few chickens are being kept and where they are enclosed in a fenced field to protect them from vermin. The chickens are usually housed inside the ark at night.

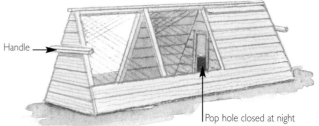

Handle

Pop hole closed at night

Walk-in shed

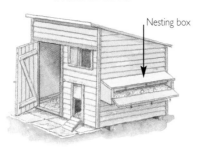

Nesting box

A traditional walk-in shed, with vents at the front, and a bank of nesting boxes. These simplify the tasks of cleaning the shed and collecting the eggs.

Hen house on wheels

Usually home-made and ideal for more chickens than an ark can accommodate. A wire-netting run can be fitted on one side. Use large wheels to ensure it can be easily moved.

Poultry shed and scratching run

Similar to a walk-in shed, but with a large, wire-mesh run attached to it (useful when housing young hens). This is ideal in areas where foxes or vermin are prevalent.

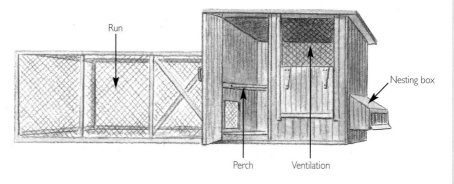

Run

Nesting box

Perch   Ventilation

A CHICKEN HOUSE AND EQUIPMENT

Roof

Chickens must be kept warm and dry to prevent the onset of diseases. Roofing felt is ideal; coat it with a roofing sealant every year. For large poultry sheds, corrugated fibre-glass roofing panels are easily fitted.

Perches

Never overcrowd perches – allow at least 20 cm (8 in) of space for each bird. Use 7.5 cm (3 in) thick, clean timber, with upper edges rounded. Position them 60 cm (2 ft) high and in a level plane (not ranged upwards).

Ventilation

Good ventilation is essential to prevent birds suffering from respiratory diseases. Install ventilators in the enclosed area – rodent-proof, louvred types at the top and hinged ones lower down (ensure they are fox-proof).

Feeding and drinking

Dry mash, corn or grit can be given through troughs or suspended self-feeding devices. Provide water through a simple water fountain or a low drinker for young chicks (make sure it is positioned at a suitable height).

Nesting boxes

There should be one nesting box for every three hens. They are best positioned so that the hen can enter from inside the poultry house, while the egg collector can lift up the lid of the nesting box from the outside.

Floor

Essentially, it must provide a dry base. Solid concrete is ideal and prevents rats burrowing in. Rot-proof, tongue-and-grooved wood forms a level, functional floor, while well-rammed gravel and rubble produces a cheap floor.

Hoppers and large containers

These are essential for keeping dry mash, corn and grit clean and dry. They must be conveniently positioned and, preferably, in a dry shed. Ensure that lids are secure and vermin-proof.

Dust baths

A hen's natural behaviour is to take a dust bath. She will search out a natural area of fine, dry earth and wallow in it. This removes parasites such as lice. Alternatively, provide hens with a litter of wood shavings.

VERMIN PREVENTION

Strong, durable fencing is essential if foxes and other predators are not to decimate your chickens. Use strong, 2.4–3 m (8–10 ft) long, timber posts, treated with a wood preservative. Hammer them 45–60 cm (1–2 ft) into the ground every 3 m (10 ft) along the perimeter. Secure galvanized, 18-gauge, 5 cm (2 in) mesh wire-fencing to the posts, with 45–60 cm (1–2 ft) buried and 1.5–1.8 m (5–6 ft) above the ground. Use a sledge hammer or, preferably, a post driver (strong metal tube, sealed at one end, with handles on opposite sides) to drive each post vertically into the ground.

Feeding your chickens

How does digestion work?

The digestion of chickens is different from that of humans. Food, when eaten, is rapidly transferred to the gullet (oesophagus), which is a narrow muscular tube that expands to accept large, undigested pieces of food. The gullet widens into the crop, where food is lubricated on its way down the throat. Sometimes food is retained in the crop for several hours, where it is increasingly softened. It then moves into the glandular stomach and the gizzard.

Ensure that food and clean water are readily available – at an easily accessible height for chicks as well as chickens and bantams.

Hens in a group (with or without the presence of a male bird) will evolve a 'pecking order' (see page 5).

THE DIGESTIVE SYSTEM

Because a chicken's digestive system initially appears unusual, it has encouraged many questions, such as:

- **Can chickens chew?** No. They do not have teeth and therefore use their horny beaks to break up food.
- **What happens when they swallow?** Food passes directly into the gullet and into the crop. Basically, the crop is a storage area for food, which is softened before travelling on its way.
- **Do chickens have a sense of taste?** Yes. They have glandular taste buds associated with the ducts of the salivary glands.
- **Is grit essential for their digestive system?** Yes. Grit in the gizzard acts like teeth. Although food can be ground by normal muscular action in the gizzard, it is not so efficient as when grit is present, which increases the number of grinding surfaces. This enables better and fuller action of the chicken's digestive juices.

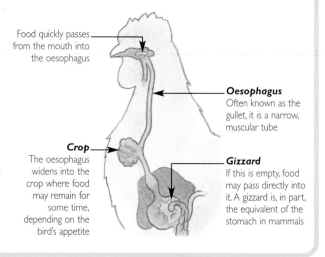

Food quickly passes from the mouth into the oesophagus

Oesophagus
Often known as the gullet, it is a narrow, muscular tube

Crop
The oesophagus widens into the crop where food may remain for some time, depending on the bird's appetite

Gizzard
If this is empty, food may pass directly into it. A gizzard is, in part, the equivalent of the stomach in mammals

NUTRITIONAL REQUIREMENTS

Apart from a supply of clean water (replenished daily), there are nutrients a chicken needs throughout its life to produce eggs regularly. The amounts and proportions of these change during a chicken's life. For example, chicks need more protein than older birds, and especially during their first few weeks.

Nutrients and their functions:

- **Proteins:** body building and repairs to damaged parts.
- **Carbohydrates:** energy.
- **Fats and oils:** energy.
- **Vitamin A:** disease resistance and balanced growth.
- **Vitamin B complex:** optimum growth development.
- **Vitamin D:** strong egg shell and healthy growth (prevention of rickets).
- **Vitamin K:** healthy blood.
- **Calcium and phosphorus:** strong egg shells and healthy bones.
- **Zinc:** feather development and healthy skin.
- **Manganese:** strong egg shells.
- **Iodine:** control of metabolism.

Kitchen scraps and grain will not provide all of these essential nutrients, and therefore a balanced feeding programme is essential.

CAN I FEED CHICKENS WITH GRASS CUTTINGS?

Yes. After mowing a lawn (untreated by weedkillers) spread the cuttings on a clean surface and allow them to dry. They can be used as a winter food supplement, as well as given fresh and green.

However, grass cuttings give the yolks of eggs a deep yellow colour, which does not suit everyone.

CAN I USE COOKED HOUSEHOLD SCRAPS?

Yes. Potatoes and other kitchen vegetable scraps when thoroughly cooked are ideal for feeding to chickens. Additionally, they have the bonus of being more easily digestible by chickens than when fed raw. But they all must be well cooked to prevent diseases spreading to your chickens.

Never accepts gifts of partially cooked kitchen scraps from friends and neighbours, as you may not know all their content.

WHAT ARE MASH, PELLETS AND CRUMBS?

- **Mash** describes any balanced food mixture. Where poultry is kept intensively, dry mash is often used because the process of eating keeps birds busy – usually when they have nothing else to do. Wet mash, however, includes vegetable waste from kitchens (together with dry mash), all mixed with hot water. Incidentally, adding water to dry mash does not improve its nutritional value, nor its appeal to the birds.
- **Pellets** are foods that have been ground and compressed into cylindrical form. They are easy to use and handle; their desired size depends on the age of the poultry. For easy feeding, the pellets can be given through self-feeding hoppers. This is ideal if you are away from home all day.
- **Crumbs** are food given in granular form.

Wild fruits and nuts

Many nuts and fruits are ideal for feeding to chickens. These include:

- Acorns: from species of oak trees (*Quercus* spp.) – dried and crushed.
- Beech nuts (mast): from *Fagus sylvatica* – dried and crushed.
- Hedgerow fruits: blackberries – fresh and in small quantities).
- Horse chestnuts: from *Aesculus hippocastanum* – dried and crushed.

- Rowan: from *Sorbus* spp. – fresh and in small quantities.
- Sweet chestnuts: from *Castanea sativa* – dried and crushed.

Home-grown vegetables

This is an inexpensive way to feed chickens and includes:

- Beans – dwarf, French, runner types.
- Cabbages and other brassicas, such as Brussels sprouts, kale, broccoli and calabrese.
- Carrots
- Jerusalem artichokes
- Lettuce
- Parsley
- Potatoes (do not use when green or sprouting)
- Turnips and swedes

Wild plants

These are widespread and can be found in your garden, along hedgerows and at the edges of fields. Ensure, however, that they have not been sprayed with herbicides and other chemicals.

The following wild plants are widespread and can be found on both waste and cultivated land:

- Fat Hen (*Chenopodium album*): annual, up to 45 cm (18 in) high. Feed when fresh and green.
- Groundsel (*Senecio vulgaris*; see right): annual, 30 cm (12 in) or even 45 cm (18 in) high. Feed when fresh and green.
- Small Stinging Nettle (*Urtica urens*): annual, 10–45 cm (4–18 in) high, with characteristic nettle-like leaves. Feed young, boiled shoot tips to chickens.
- Perennial Nettle (*Urtica dioica*): Perennial, 30–90 cm (1–3 ft) high, with characteristic nettle-like leaves. Feed young, boiled shoot tips to chickens.

Handling and inspecting chickens

Most chickens are friendly and quiet and enjoy the company of poultry enthusiasts. The more they are handled, the easier it becomes. If chickens need an incentive to be friendly, first offer them a handful of food pellets. In all groups of chickens, there is a strict pecking order; if you pick up a chicken that is low down in this order, those higher up may become jealous and walk over to check what you are doing and to make their presence known.

CATCHING AND PICKING UP A CHICKEN

When trying to catch a particular chicken, quietly usher it into a corner (see above); do not frighten it by frantically and excitedly chasing it in circles. Panicking a hen will reduce its laying ability and be bad for its general health. Excessive noise and frenzied acitivity at this stage will also reduce its desire to lay eggs.

Do not grab at a chicken's wings or tail. Once the chicken is in front of you and by your feet, bend down slowly and pick it up by hugging its wings close to its body (see below). It is then not able to injure itself by suddenly flapping its wings and, perhaps, falling from arm height onto the ground, when it might become injured. Heavy breeds are especially at risk if they fall.

Once a bird is safely between your hands, draw it into your body to secure it further and reassure it that it is safe.

Chickens enjoy the attention of being picked up and stroked, and they love being talked to calmly.

Take care when handling chicks as they are easily frightened and may jump off – but they are lovely to hold.

CARRYING A CHICKEN

When the chicken is secure, put one hand under its rear end and tuck its head under your arm (see above). The head should be slightly lower than the bird's rear end. Some poultry enthusiasts like to have the chicken's head facing forward, but this does not enable its vision to be slightly restricted, thereby avoiding the chance of sudden visual shocks frightening it. The bird must be kept calm.

SUPPORTING AND CHECKING A CHICKEN

To hold the chicken firm and confidently while checking it, slip your hand under the bird, with the middle two fingers between the legs (see above). It is then possible to check it and give support.

At what time of day should I inspect my chickens?

Preferably, this is best undertaken at dusk, but when still in good light. Hens will have settled on their perches and you will be able to see the state of their crops; if they are full, you are giving them enough food, but if empty more food is needed.

Generally, the bird's body should be firm to the feel, without undue heaviness or flabbiness. If the bird feels slightly on the 'spare' side, this is probably the result of her egg-laying activities. Check her crop to see if it is full; if not, make a note to increase her food.

THINGS TO CHECK

By catching and lifting up a chicken in good light, rather than waiting until dusk, it is easier to inspect a chicken in detail, although take care not to alarm the bird. However, checking a bird's crop is best undertaken at dusk, after she has had a meal (see box, below left).

- **External parasites** (see pages 62–63) may be present. Draw the feathers open to check their undersides, as this may be where they are congregating. Also inspect areas of bare skin. If lice, ticks, fleas and mites are allowed to reach epidemic proportions, they are extremely difficult to eradicate.

- **Broodiness** (a bird with a desire to sit on eggs and hatch them) can be checked by lifting up the tail – a hen in this state shows loss of feathers from its breast (see above).

- **Whether a bird is laying** is easily assessed – if you can place four fingers between the pelvic bones and the end of the breastbone, the hen is in full lay. However, if there is only space for two of your fingers, the bird is not laying (see above).

TWO IMPORTANT DON'TS

- **Don't** pick up a chicken by its feet or neck, as this causes stress and damage.
- **Don't** pretend to hypnotize a chicken by placing it on its back, as this may lead to the bird suffering heart failure (especially if a heavy breed).

Chickens for the table

Can I eat egg-laying chickens?

When a hen starts to lose the ability to produce eggs, it often becomes necessary to make the decision to kill her humanely (see pages 40–41). However fond you may have become of a particular hen, when it costs more to feed and keep her than the value of her eggs, it is time for her to go. If not destined for the table, the carcass must be disposed of in an environmentally friendly way, and never buried on your land (see below for details).

Disposing of a dead chicken

If, after killing a chicken, you decide not to eat it, the body must be disposed of legally. This also applies to chickens killed in road accidents, and those that die through old age or have been euthanased because of illness. Immediately you have a dead chicken and wish to dispose of it, contact your local Animal Health Office, an organization that gives advice about dealing with dead animals. In most areas, the organization that can assist you is the National Fallen Stock Company (view their website for detailed information).

THE CYCLE OF LAYING AND EATING

Because the egg-laying ability of hens kept primarily for their eggs diminishes after their second laying season, eventually they are best killed and eaten. The flesh on such hens will be tougher than on young birds which have been fed and kept solely for meat production. However, this need not be an insurmountable problem as an old hen can be killed, plucked and prepared for cooking (see pages 40–45) and simmered in hot water for several hours before being placed in an oven. This helps to tenderize the flesh.

WHAT IS A CAPON?

This is a term for a male chicken that has been neutered by surgery or chemicals to shorten its growing and fattening time and to improve its eating quality. If you breed your own chicks, about half of them will be male – the best way to treat them is to have them neutered and later use them for eating. They undoubtedly grow and fatten more quickly if they are caponized.

Traditionally, caponizing involved making an incision in the bird's side – between the last two ribs and closest to the tail – and removing the testicles. This cutting and removal or reproductive parts involved a certain amount of cruelty, but nowadays caponizing a bird can be achieved chemically. Female hormone oestrogen is inserted in pellet form under skin in the bird's neck. It results in a shrinking of the testicles and reversing of the bird's sexual characteristics.

However, both surgically caponizing a male chicken and putting female hormone oestrogen as a pellet in the bird's neck have been banned in many countries – the first method because it causes unnecessary distress to a bird, and the second because of the risk of female hormones being absorbed when the bird is eaten.

Eventually a decision has to be made about humanely killing a chicken, and perhaps eating her. This can be a very difficult decision if your chickens have been treated as family pets, rather than just as egg producers.

Alternatives to surgery and the use of hormones

• Separate male birds from the females when 12–14 weeks old and place them in a small pen. Do not allow them to roam freely. Three times a day, feed them with a mixture of equal parts oats, barley and boiled potatoes, with the addition of a little skimmed milk. Give the birds as much of this mixture as they can eat, as well as fresh water. They fatten quickly and can be killed and eaten about three weeks later.

• Another – and more modern method – is to buy young, hybrid birds that have been specially bred for the production of meat for the table.

BREEDS FOR MEAT AND EGGS

Meat (table breeds)

*Croad Langshan
(see page 15)*

Dorking (see page 20)

*Rhode Island Red
(see page 21)*

Colour varieties

Some breeds have several varieties. For example, the Wyandotte has four that are described and illustrated in this book – all are useful to the home poultry keeper.

Some breeds, however, have colour varieties that are better for eating. The white form of the Plymouth Rock (see below) is one example.

Eggs and meat
(general-purpose breeds)

*Barred Rock
(see page 11)*

*Black Australorp
(see page 11)*

*Buff Orpington
(see page 14)*

*Light Sussex
(see page 18)*

*Maran
(see page 16)*

*Plymouth Rock
(see page 20)*

*Rhode Island Red
(see page 21)*

*Wyandotte
(see pages 13, 17, 21 and 22)*

Killing chickens

Is killing a chicken difficult?

The act of killing a chicken is easy, but experience is essential to ensure that it is despatched humanely and without any unnecessary suffering. There are animal welfare standards to consider and these must be complied with at all times (see below for details). Of the several ways to kill a chicken, the most practical and popular one used by home poultry enthusiasts is described and illustrated on the opposite page.

The cycle of birth, life and death is always present when keeping chickens, but a humane end to their lives is essential. Causing unnecessary distress to a chicken is both immoral and illegal.

Can I legally kill chickens?

The legalities of killing home-kept chickens humanely varies from one country to another, but basically they are:

Can I eat meat from my own chickens?
Yes – but you must ensure they are slaughtered lawfully.

Where can I slaughter my own chickens?
The options are to kill your own chickens on your own land, or to arrange for them to be killed in an approved slaughterhouse. It is unlawful to have your chickens killed anywhere else.

Can I sell meat from my own chickens?
No – this would be illegal. Only if the chicken was killed in an approved slaughterhouse would it be possible to sell the meat or even to use it to feed paying guests in a bed-and-breakfast enterprise owned by you.

METHODS OF KILLING CHICKENS

There are several ways to kill chickens and with all of them the slaughter must be carried out humanely. On a commercial scale this is achieved by methods known as 'electrical stunning', 'controlled atmosphere stunning' and 'concussion stunning', where a captive-bolt instrument is used. A variation on this method is a weighted stick, although this is not recommended.

For home poultry keepers, the traditional way to kill a chicken has been 'neck dislocation', also known as 'neck pulling', and it is still the best way to despatch an injured chicken or one that is going to be eaten. If you are anxious about the procedures involved and getting it right so that the chicken experiences the least stress, consult with a local chicken-keeping club or enquire about a 'poultry course' in your area, which both demonstrates and explains the techniques of slaughtering chickens. Killing a chicken is not for a person with a sensitive stomach, nor for anyone without experience in undertaking this role.

Basically, this method of killing a chicken involves dislocating the neck, causing concussion and rupturing the spine. When done correctly, the bird immediately and irrecoverably loses consciousness.

HOW TO KILL A CHICKEN

Take care not to stress the chicken that is to be killed, nor the remaining chickens. Select the chicken and take it to another enclosure or, preferably, an empty shed so that the bird is on its own with you. Putting it in a large box covered with a cloth helps to ensure it remains calm. Alternatively, provide the bird with a perch that is about 30 cm (12 in) high.

How the chicken is held is influenced by its size and the despatcher's height. If you are a short person, standing on a firm, slightly raised surface so that the bird hangs freely is a solution. However, where possible just standing on level ground is the best way as this gives the greatest stability and ensures that you will not inadvertently step backwards and fall over.

Step-by-step to killing a chicken

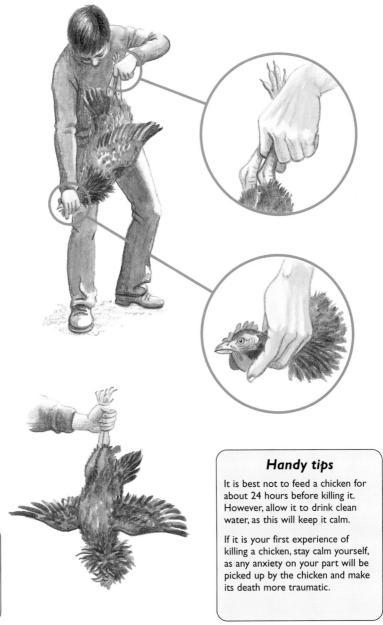

1 *Hold the chicken's legs upside down with your left hand. Grasp the legs with your thumb pointing away from your body; this enables a firm and strong grip to be used (see inset). Ensure that the chicken's head is hanging across the front of your body and not away to the opposite side.*

2 *With your right hand, hold the chicken's neck so that three fingers are to one side and holding the neck horizontally. The thumb should tightly clasp around the other side of the neck (see inset).*

3 *With a sudden pull downwards, dislocate the chicken's head by snapping it and leaving a gap between the neck and the last vertebra. It is better to pull downwards too firmly than not to break the neck. If the head comes away from the chicken, this can be messy but it is better than not completing neck dislocation.*

4 *The bird usually flutters its wings and moves its legs for a few moments; this is quite natural, but the bird will be dead. During this period, ensure that the bird is still held firmly.*

5 *The bird can then be suspended by its feet from a strong hook in a cool, shaded place.*

Left-handed?

If you are left-handed, reverse the positions of the hands, so that there is a strong hand for pulling down the chicken's neck.

Handy tips

It is best not to feed a chicken for about 24 hours before killing it. However, allow it to drink clean water, as this will keep it calm.

If it is your first experience of killing a chicken, stay calm yourself, as any anxiety on your part will be picked up by the chicken and make its death more traumatic.

Plucking chickens

I t is essential to remove all of the feathers from a dead chicken, either immediately after death or when the body is cold (see below for details). The removal of feathers enables the chicken's exterior to be cleaned and dried, as well as preparing it for the other pre-cooking tasks such as hanging, drawing and trussing. Each of these jobs is vital in the preparation of a chicken for eating, and should always be carried out carefully and thoroughly.

HOW SOON AFTER DEATH SHOULD I PLUCK A CHICKEN?

The body of a chicken will remain warm and floppy for about 45 minutes after death and before rigor mortis sets in. Some poultry enthusiasts prefer to pluck a chicken before it is cold, when the feathers usually come out easily.

The time of plucking is therefore a matter of preference, but if a chicken cannot be plucked immediately after death leave it until quite cold. Do not pluck a chicken when half cold as the skin will be torn and damaged.

SCALDING A CHICKEN

This technique is recommended by some poultry enthusiasts to aid the removal of feathers. It involves dipping the chicken in hot water (too hot to allow a hand to be immersed). A more precise guide is 62–65°C (145–150°F). The hot water melts the fat around the bases of feathers and enables them to be easily pulled out.

The method of scalding is to hold a chicken by its feet and to dip the body (and legs) into the water for 3–5 seconds, at the same time jiggling it to ensure that water penetrates the bases of the feathers. Then remove the chicken and after a few seconds replace it in the water, again for 3–5 seconds.

A chicken that has led a free-ranging and happy life should produce a tasty bird for the table.

Remove the bird and tug at a large wing feather; if it comes out easily, the scalding is complete. If not, re-immerse the chicken and again check if a feather comes out easily. This may need to be repeated several times.

WHAT DO I DO WITH THE FEATHERS?

These are high in nitrogen and ideal for adding in thin layers to a compost heap. Small, young feathers decay faster than old ones with thick quills (the stiff and often thick centre). Therefore, you should not be surprised if after a year of decomposition parts of them are still recognizable. This is seldom a problem, however.

The resulting nitrogen-rich compost is ideal for digging into the soil when preparing planting or sowing positions for runner and French beans, as well as other nitrogen-loving vegetables (see pages 30–31).

The practicalities of hanging a chicken:

Birds are hung for 24 hours to reduce the core temperature to 0°C (32°F). This is best done in a cool, clean, vermin-proof place with a good circulation of clean air and a temperature under 3°C (37°F).

Alternatively, place the chicken in a refrigerator. Tie a plastic bag over the chicken's head to catch drips of blood from the nostrils and beak.

Do I need to hang a chicken?

The decision to hang a chicken is a matter of personal choice. The advantages of hanging are:

- The flavour of the meat is improved.
- The texture of the meat improves with the onset of its natural breakdown.
- The viscera (soft tissue within the chicken) firms up and aids its removal (see pages 44–45).
- Fats in the chicken's body solidify.

PLUCKING A CHICKEN

Wear an overall to ensure that your clothing does not become spoiled. A face mask is also often necessary when you are working in fully enclosed conditions, where dust might be raised during feather removal.

- Sit on a low stool, with the chicken's head downwards and across your lap. Alternatively, suspend the bird by its legs from a strong hook or other secure fixing positioned level with your head; this usually ensures that most of the chicken's body is level with your shoulders, making it easy to reach and remove the feathers.
- It is essential that the feathers are pulled out from the direction in which they grow, rather than twisting them and pulling at an angle.
- Either place a large dust sheet underneath the bird, or carry out the plucking on a clean, flat surface, so that the feathers can be easily collected and brushed up.

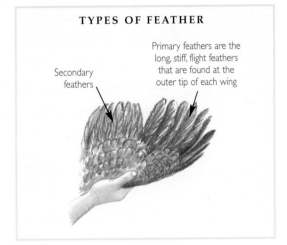

TYPES OF FEATHER

Secondary feathers

Primary feathers are the long, stiff, flight feathers that are found at the outer tip of each wing

1 *Start with the primary feathers. Holding several feathers firmly together, give them a sharp tug. This avoids tearing the skin.*

2 *Pluck the leg feathers next.*

3 *The body is the next area to be plucked.*

4 *Eventually, the bird's body and legs will be free from feathers. The neck can then be cut off at the point of dislocation, and the bird 'drawn' (see pages 44–45).*

Drawing and trussing a chicken

Are drawing and trussing necessary?

It is essential to 'draw' a chicken, which means the removal of all of its 'innards'. Sometimes, this is known as 'evisceration' or 'gutting'. Parts of the innards, such as the heart, the liver and the outer case of the gizzard (the part of the stomach where the bird grinds its food), as well as the remainder of the neck, are known as the 'giblets' and are occasionally retained for simmering into a nourishing and traditional broth.

DRAWING A CHICKEN

Before tackling this task, ensure that the table is clean and provision is made for the collection of the chicken's innards. This is essential as you will not want to search for a bowl when your hands are covered with pieces of intestine!

Cleaning the gizzard

Before setting the gizzard to one side – perhaps ready for cooking – it needs to be cleaned. Use a sharp knife to cut halfway through it, then pull out the yellow bag of grit; this should be discarded.

1 *Position the chicken on a flat, clean, washable surface. Hold the body firmly and use a sharp knife to cut the neck to about 2.5 cm (1 in) above the bird's shoulders. Slit the neck further, cutting through the muscle so that it breaks away.*

2 *Peel away the crop (the portion of the chicken's food channel, where food is stored until it is ready for passing into the stomach), and cut it out close to the neck cavity. Insert a finger into the neck cavity (see above) to loosen the ligaments close to the breastbone; they then can be prised away and removed.*

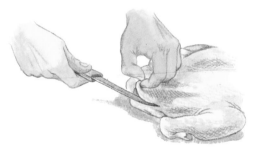

3 *Turn the bird around, so that its rear end faces you. Use a sharp knife to form a wedge cut to sever the fatty oil sac (sometimes known as the 'Parson's nose') on the rear end of the bird and positioned close to its tail. If left, it sometimes gives a peculiar flavour to the meat.*

4 *Make a cut between the Parson's nose and the vent (the orifices through which the chicken defecates, urinates and expels eggs). Carefully cut around this – taking care not to cut the gut – and pull out the intestines.*

DRAWING A CHICKEN CONTINUED

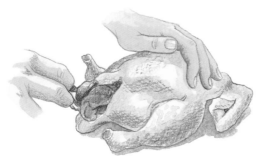

5 *Continue to remove the innards, which include the heart, liver, viscera and kidneys. Then check to ensure that all of the inside of the chicken has been removed. Lightly wash and thoroughly dry the inside, taking care not to excessively wet it. Sometimes, all that is necessary is to use a damp cloth. Bacteria that might be present are killed during cooking. However, always ensure that the cooking process is thorough, at the correct temperature and for the recommended time.*

Singeing the carcass

If there are a lot of small hairs on a chicken's body, these can be quickly removed by singeing. Move a burning piece of paper over the hairs, or hold the carcass over a lighted gas jet. But take care, as the skin should not become burned.

Chicken preparation courses

If you are in doubt about your ability to prepare a chicken for eating, local food preparation and home cookery courses are well worth attending. As well as textbook advice, there is often the opportunity of getting some practical hands-on experience.

TRUSSING A CHICKEN

Trussing is often considered to be part of drawing, but to many poultry enthusiasts it is a separate task and another step in preparing the bird for cooking.

1 *Fold loose pieces of neck skin back and over the resulting hole. Place the bird on its back. Pass a piece of strong thread across the thighs and above the hocks. Then take the ends of the string up and alongside the body and tie them firmly.*

2 *Use another piece of string to secure the legs. Sometimes a single piece of thread is used to hold both the legs and the thighs in place.*

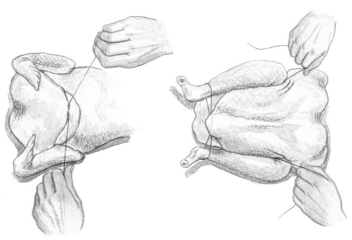

FREEZING, DEFROSTING AND STORING

Once a chicken has been drawn and had its innards removed, it needs to be cooked as soon as possible. Alternatively, it can be placed in a home deep-freezer (but not if it has been 'stuffed'). Place it in a moisture-resistant wrapper, with the giblets in a separate bag.

Essentially, the deep-freezer should be set to its coldest, so that the chicken loses temperature rapidly. This helps to ensure that there are no internal changes to the chicken and that it retains its natural flavour and appearance. A temperature of −18°C to −34°C (−0.4°F to −29.2°F) is essential.

Defrost a chicken slowly. Place it in a refrigerator for a couple of days; once thawed, it can be left there for a further day or two before being cooked.

Alternative ways to defrost a chicken include in cold water and in a microwave, but placing it in a refrigerator is the best and safest method.

Chickens defrosted in a microwave must be cooked immediately afterwards. This is because parts of the chicken may have become warm and begun to cook; if there is an interval between defrosting and cooking any bacteria present might multiply and cause food poisoning.

Eggs for the home

What is involved?

Keeping hens in a poultry house in your back garden or yard to provide you with eggs becomes a way of life (see pages 6–7 for the regulations involved in keeping chickens). Here is all you need to know about the production of eggs, their sizes and structure, and the cycle of egg production, as well as how to encourage pullets (female chickens less than one year old) to start laying eggs. Where possible, it is worth keeping a daily record of eggs produced.

HOW MANY EGGS CAN I EXPECT A HEN TO PRODUCE EACH YEAR?

This very much depends on the breed of chicken and if it is classified as a large fowl or a bantam. In the descriptions of breeds of chickens and bantams (pages 10–27), the expected number of eggs for each breed is indicated. Some breeds, such as the Silkie, produce 80–100 eggs in a year, while others have been specially bred for their high egg-laying abilities. These include the Black Rock (280 or more eggs in a year). Bantams lay fewer eggs.

AT WHAT AGE WILL PULLETS START TO PRODUCE EGGS?

Pullets usually start laying eggs at about the age of 20–22 weeks, sometimes slightly earlier and occasionally longer. Often, pullets when first laying have problems in passing an egg. This may be a result of egg binding – the cause and treatment are explained on page 67. Other problem habits with chickens are described on pages 66–67.

Black Rock

Regular clutches of eggs are the reward for keeping your hens properly fed and watered, and comfortably housed.

Why do we eat eggs?

A chicken's egg is one of the miracles of nature. It is primarily intended for reproduction, but has been hijacked by millions of people each day as a food that provides an invaluable source of protein. Chickens are so good at feeding us that currently it is estimated there are more than 28 billion of them! Medical research also suggests that eating eggs may prevent age-related macular degeneration of the eyes.

How many calories does an egg provide?

Depending on its size, each egg provides 60–80 calories, but when in a fried egg sandwich – and together with butter and bread – this rises to about 225 calories. Incidentally, militarily a fried egg sandwich is known as an 'egg banjo'. However, whether on their own at breakfast time, in a sandwich or formed into an omelette, eggs are by far the best fast food and certainly more healthy than many processed fast foods. They also create less expensive meals.

Are they useful when on a diet?

Yes. This is because in order to lose weight the number of calories you consume must be reduced, while maintaining a balanced and nutritional diet.

How much does an egg weigh?

An average-sized egg weighs about 57 g (2 oz); the shell forms 11% of this, the white about 58% and the yolk 31%.

WHAT IS AN EGG?

An egg is formed of many parts and all within a superbly shaped and engineered shell, well able to withstand the stresses incurred when being expelled by a hen. It is also able to resist knocks and damage. Within an egg there are several parts – the white and yellow elements are widely known, but there are others, as shown below:

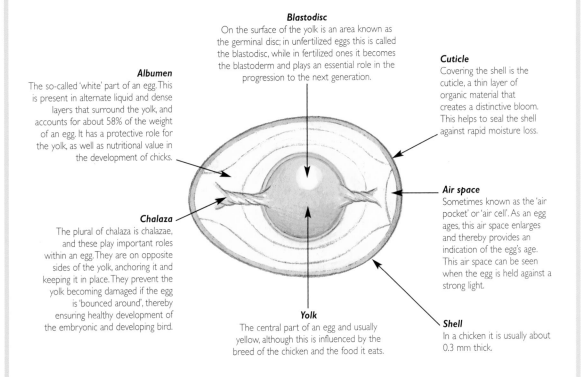

Blastodisc
On the surface of the yolk is an area known as the germinal disc; in unfertilized eggs this is called the blastodisc, while in fertilized ones it becomes the blastoderm and plays an essential role in the progression to the next generation.

Cuticle
Covering the shell is the cuticle, a thin layer of organic material that creates a distinctive bloom. This helps to seal the shell against rapid moisture loss.

Albumen
The so-called 'white' part of an egg. This is present in alternate liquid and dense layers that surround the yolk, and accounts for about 58% of the weight of an egg. It has a protective role for the yolk, as well as nutritional value in the development of chicks.

Chalaza
The plural of chalaza is chalazae, and these play important roles within an egg. They are on opposite sides of the yolk, anchoring it and keeping it in place. They prevent the yolk becoming damaged if the egg is 'bounced around', thereby ensuring healthy development of the embryonic and developing bird.

Air space
Sometimes known as the 'air pocket' or 'air cell'. As an egg ages, this air space enlarges and thereby provides an indication of the egg's age. This air space can be seen when the egg is held against a strong light.

Yolk
The central part of an egg and usually yellow, although this is influenced by the breed of the chicken and the food it eats.

Shell
In a chicken it is usually about 0.3 mm thick.

WHAT ARE THE SIGNS OF A PULLET STARTING TO LAY EGGS?

An indication that a pullet is starting to lay eggs is the display of a bright red and full comb. You will also notice a widening – sometimes known as 'filling out' – of the pullet's body and especially on either side of the 'vent' (the orifice from where a chicken lays an egg). This area can be inspected to assess if a chicken is about to begin laying (see pages 36–37 for handling and inspecting a chicken, and judging if she is able to lay eggs).

ENCOURAGING A RELUCTANT PULLET TO LAY

A pullet that is reaching a stage when she might begin laying eggs will display restlessness, moving in and out of the hen house and clucking in a disturbed and dominant manner. If she is in with a group of chickens that are established layers, there will be several nesting boxes already present for them. If not, ensure that one is put in a dark and quiet corner of the house, and place straw inside it to create a reassuring nesting area for her.

Occasionally a pullet will lay eggs on the floor; quickly remove them, together with surrounding and contaminated straw. Placing replica china eggs in a nesting box will encourage her to lay eggs in the correct place.

CAN I SELL THE EGGS MY HENS PRODUCE?

The rules and regulations of selling spare eggs to friends or through local shops is decided by national or regional health and welfare agencies. In the British Isles this is the Department of Food and Rural Affairs (DEFRA). In Europe, the sale of eggs must be compliant with the EU Welfare of Laying Hens Directive.

This means that, although it is not impossible to sell your spare eggs, be prepared for several visits from ministry officials. Apparently, all Class A eggs sold at retail level within the EU are legally required to be marked with a code identifying the egg-laying establishment, country of origin and method of production (organic, free-range, barn or cage). Ungraded eggs sold direct to the final consumer are exempt from these markings, although those sold through local markets, either graded or ungraded, have to be marked.

Fertilizing and collecting eggs

Is a cock bird essential?

The presence of a male bird (known as a cock bird, cock or rooster) is not essential for hens to produce eggs. However, if you want to breed from your hens, a cock bird is vital. There are both benefits and problems when having a cock bird present in your group of hens and these factors are outlined below. However, for many poultry enthusiasts the sight of a rooster strutting among his 'ladies' creates a countrified and idyllic picture.

ADVANTAGES AND DISADVANTAGES OF KEEPING A COCK BIRD

To many poultry enthusiasts, whether or not to have a cock bird is a balance between being able to breed your own chicks and the cost and noise involved in keeping one. Additionally, there is no evidence that the presence of a cock bird will encourage your hens to produce more eggs. Also, the introduction of a cock bird to hens who have not seen or heard one before can be disturbing for them and may even put them off laying eggs. Below are a few considerations about keeping a cock bird:

Advantages

- He fertilizes your hens and they then produce eggs for hatching and the subsequent production of chicks.

- The number of 'fertile' eggs produced daily by your hens increases with the added time he is with them.

- He will keep your hens under control, preventing squabbles breaking out between them.

- His presence helps to keep hens placid and contented.

Disadvantages

- Noise – especially in a suburban garden and where neighbours may be quite close.

- Cost of buying.

- Cost of feeding.

- He needs to be ten or more months old.

- Fertility declines after he is 5–6 years old.

- Check that he is not closely related to the hens, as this may result in in-breeding and chick deformities.

- He should have been vaccinated against Fowl Pest (see page 64). This is also known as Newcastle Disease, and is notifiable (see page 61).

- He should not have more than eight hens to look after, but preferably only six.

- Buy him from a reputable source.

- Do not put a cock bird in with young pullets if you want good hatching eggs.

Pride, importance and majesty are all self-evident in a cock bird. In some countries he is better known as a rooster.

Collect eggs regularly each day (see opposite page for details), placing them in a deep basket or bowl.

COLLECTING EGGS

Commercial egg producers remove eggs as soon as possible after they have been laid, which can be 3–4 times a day. Home poultry keepers – especially if occupied on other jobs – usually just gather them from nesting boxes in the morning and evening. The benefit of gathering your eggs as soon as they are laid is that it gives a hen very little opportunity to peck at them.

EATING FRESHLY LAID EGGS

Many poultry enthusiasts like to leave newly laid eggs for a couple of days before eating them; this allows the white parts to settle. To ensure that you know which eggs are more recent, lightly write on them with the date as soon as they are collected.

CAN I STORE EGGS?

Fresh eggs can be stored for 2–3 weeks in a cool, well-ventilated larder, while if they are kept in a refrigerator at a constant 4°C (39.2°F) they will remain edible for 3–4 weeks. Eggs cannot be stored in a freezer because their shells would burst on freezing.

Whole yolks remain edible in a cool larder for a day, and 4–5 days in a refrigerator; the whites of eggs will keep for a week in a refrigerator. However, it is always best to store eggs in their entirety and to use them fresh. If stored as yolks or white parts, place them in clean, clear bowls and cover these with clingfilm.

TESTING AN EGG FOR FRESHNESS

This is a straightforward and quick procedure. Put the eggs in a bowl of water: any bad ones will float, while the fresh ones sink to the bottom of the bowl. This is because the air space (see page 47) is smaller in a fresh egg than in an older one.

EATING FERTILE EGGS

These are eggs that come from a hen that has been fertilized by a cock bird. They are all right to eat, although some people do not approve of them. Registered and commercial producers of eggs are sometimes required through egg marketing regulations to produce eggs with a 'yolk that is free of foreign bodies', which means that it has not been fertilized.

BROODINESS – WHAT IS IT AND HOW DO I PREVENT IT?

Broodiness is when a hen insists on sitting on eggs rather than laying them. This trait can be discouraged by putting her in a slatted coop (see above right) by herself but within sight of other hens. The coolness of this coop, together with a few days on her own and watching the other hens scratching at the soil and apparently enjoying themselves, usually discourages her from her broodiness. She can then be put back with the other chickens to lay further eggs with no problems.

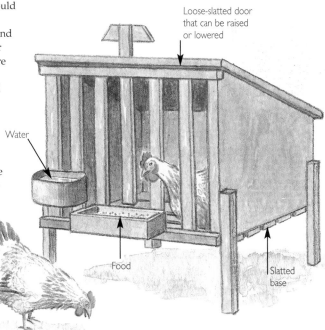

Loose-slatted door that can be raised or lowered

Water

Food

Slatted base

Abnormal eggs

These are eggs that show unusual characteristics, including:

Double yolks: This is not a problem and often occurs in hens in their second season of laying. Some breeds of hens are prone to this and it is also thought to be hereditary.

Wind eggs: Eggs which do not contain a yolk. This often occurs in young pullets which are not fully developed. Conversely, it may happen in hens which are coming to the ends of their egg-laying life.

Eggs with blood spots: They are quite harmless and thought to be a hereditary problem. Eggs can be 'candled' (see page 57) and the problem ones removed.

Soft-shelled eggs: This occurs when there is a deficiency of calcium in a hen's diet. Therefore, add calcium to their food. Loud noises also result in the production of soft shells.

EGGS IN WINTER

Chickens lay fewer eggs in winter than in summer. This is because they are stimulated into egg production by increasing daylight. It also answers the question about why there is an increase in egg production in spring.

Commercial egg-producing establishments provide their birds with artificial lighting that helps to maintain egg production throughout the year. Home poultry keepers do not usually do this, but by orientating the chicken house appropriately (as recommended on page 29) the hens can be encouraged to increase their egg laying at the earliest possible time in spring.

How to breed chickens

Is it worth breeding your own chickens?

Apart from saving money on buying young pullets that will replace aged hens at the end of their laying period, there is always the fun of having chicks running around. You will, however, need to get equipped for this influx of young life into your world. You can expect to be preoccupied with their needs for many weeks – it is a maternity unit for chickens! Although breeding chickens is time-demanding, it will interest your whole family, especially the children.

A cock bird (rooster) marshalling his 'ladies' is an integral part of breeding your own chickens.

REPLACING AGED HENS

This can be undertaken in two ways:
• Breed replacements from your existing hens (see below).
• Buy young chicks that you can nurture until they become pullets and are ready for laying (this process usually takes 20–22 weeks).
• Buy young, point-of-lay (often known as POL) pullets that will replace your aged hens. This is an easy and efficient method of replacement – the old hens can be killed (see pages 40–41) and eaten. If they cannot be eaten, the bird has to be disposed of in a legal manner (see page 38 for disposing of a dead chicken), not just buried on your land.

HENS FOR BREEDING

It is possible, of course, to use any of your hens for breeding. However, it is essential that the hen (see the panel on the right for desired characteristics) should not be closely related to the cock bird, as this may result in 'in-breeding' problems with the ensuing progeny.

Incidentally, it is far better for the home poultry enthusiast to concentrate on breeding from pure-bred breeds, rather than trying to produce your own variations on a breed.

HEN CHARACTERISTICS

Before selecting a hen for breeding purposes, check her over with a clinical eye, not an emotional one. It is no good breeding from a chicken just because she has become a family pet or has some endearing characteristics. Remember that it costs just as much to keep a hen with a poor laying record, which answers to her name and appears to recognize you, as one that is good at producing eggs.

Here are a few considerations:

• **Good laying ability:** Keep a record of the egg-laying pattern of your hens. This helps later when choosing a hen to breed from.

• **Egg quality:** Avoid hens that lay misshapen eggs.

• **Should be at least the age of sexual maturity:** This is about 22 weeks for a pullet. If a pullet does not begin laying until much later, she is best not considered for breeding purposes. Preferably, however, a hen should be in her second year of laying before she is used for breeding purposes.

• **Only breed from healthy hens:** A range of parasitic pests, diseases and problem habits is described on pages 60–67.

HOW MANY HENS CAN A COCK BIRD LOOK AFTER?

A heavy-breed cock bird should have no more than eight hens to look after – preferably six. However, a light-breed cock bird can be safely allowed ten hens.

Whatever the breed, the cock bird will regularly need a rest from the hens. You should place him in a separate pen for 1–2 days each week.

HOW LONG WILL A COCK BIRD REMAIN SEXUALLY ACTIVE AND FUNCTIONAL?

It depends on his constitution, but usually four years is the limit; however, some are known to be still active and sexually functional at 5–6 years of age.

Looking after your cock bird

- **Feeding a cock bird:** A high-protein diet is essential. Apart from his normal rations, chopped liver usually peps him up.

- **Trimming his spurs:** Keep the ends of his spurs trimmed to about 12 mm (½ in) long to prevent him damaging a hen during mating. When trimming his spurs, wrap him in a towel and use wire-cutters to cut his nails.

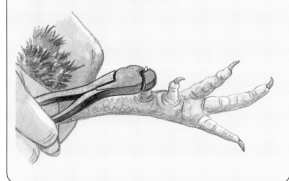

INCUBATION AND REARING

Once a hen has been mated a few times she will start to lay fertile eggs. The eggs can be left under the hen's body – where the temperature is usually 37.7°C (100°F) – to hatch out naturally.

Hens that become broody and willing to sit on a clutch of eggs can be recognized by their attitude:
- A broody hen will decide to remain on the eggs.
- If approached, she will fluff up her feathers and produce loud, guttural squawks. This is a defensive attitude, telling you to go away.
- If removed, she will walk around, still fluffed up, and make her way back to the eggs.
- Eggs hatched under a brooding hen take about 21 days before a chick appears. This time may vary slightly by one day earlier or one day later, but the hatching of eggs cannot be rushed; it is all part of their long-established reproductive cycle.

COCK BIRD CHARACTERISTICS

- If you are buying a cock bird for breeding purposes, check with the records of his parents to ensure he will produce good hens for you. It is essential that the cock bird's mother came from good stock, with a proven record of laying. Always buy from a good, reputable supplier.

- Ensure he is in good health. The general health points in a hen are featured on page 60. Many of these desired qualities also apply to a cock bird (rooster).

- He needs to be at least ten months old, and preferably older.

- He should not be related to the hens.

Unless the cock bird is healthy and confident, he will not be able to produce fertile eggs for hatching chicks.

WHAT IS THE BROODING CYCLE?

In nature, chickens lay two clutches of eggs each year, which they naturally sit on in order to produce chicks. The first clutch is in spring, the second during late summer.

The hybridization of chickens has almost eliminated this natural inclination in some breeds, but it is a characteristic that still remains in the older, pure-bred types. Therefore, if you need a hen to act as a surrogate mother to hatch eggs it is to these older breeds that you must look for assistance.

A description of a broody hen is given on page 51 under 'Incubation and rearing'.

WHERE SHOULD I PUT MY BROODY HEN?

She needs to be given special consideration, as follows:
- Provide her with gentle warmth and cosiness. The area must also be dry, free from draughts and well secured against vermin, especially rats.
- Provide her with warm and dry bedding, such as straw or hay; wood shavings are another possibility.
- She needs to be quiet and on her own, away from other hens who might intrude upon her private area.
- Check to ensure she is free from lice and mite attack (see pages 62–63 for details of the parasites).
- A nearby supply of food and clean water is essential.
- She usually gets up and takes some exercise each day. This is a chance for you to check the eggs.
- If she does not get off the eggs, gently lift her up and check the eggs are all there.
- It is not necessary to turn the eggs, as the hen will attend to this several times each day.
- During the last three days before hatching, lift off the hen and sprinkle the eggs with slightly warm water to ensure they are not becoming dry.

Keep your expectant mum in a dry, warm and cosy place, away from intrusive other hens which might disturb her.

ON HATCHING ...

A fertilized egg takes about 21 days to hatch; towards the end of this period you will notice the egg being pecked from the inside. Do not assist a newly hatching chick to escape from the shell. Those chicks that do not manage to get out of the shell are weak and will not survive for long. It may appear to be a touch of 'jungle lore', but an infirm chick will never produce good eggs in the quantity you want.

There are occasions, however, when a shell is thick and resists widening by the chick to enable it to break free. In this instance, carefully soften the shell using slightly warm water; then replace it among the other hatching eggs. Do not use water excessively as you may inadvertently drown the chick.

It is a magical time when new life appears, with a young chick trying to break free from its egg shell.

LOOKING AFTER CHICKS

Few hearts are not warmed and excited by the sight of a young, fluffy chick.

Mum and her brood of clustering and tumbling chicks form a memorable picture.

Looking after chicks

Like all youngsters, chicks need special attention and feeding to ensure they grow rapidly, strongly and healthily.

Initial feeding

For the first 24 hours the chicks will not need any kind of feeding. Thereafter, they can be fed a high-protein proprietary chick feed, usually formed of crumbs and known as 'starter crumbs', which contain all the nutrition they need for a balanced diet.

Initially, these starter crumbs are more digestible to the chicks if slightly dampened; a few drops of cod liver oil can also be added.

An alternative to starter crumbs is to feed them a mixture of a finely chopped hard-boiled egg mixed with breadcrumbs which have been finely grated from wholemeal bread; this takes longer to prepare and therefore many home poultry enthusiasts use the starter crumbs, which can be fed until the chicks are 6–8 weeks old.

During this early feeding in their lives, your chicks need to be fed regularly and often during the daytime, usually every four hours. The mother hen will show the chicks how to peck at the starter crumbs, but if they are slow to learn fill a shallow dish with food and tap its sides to attract their attention.

In addition to the food, make sure a shallow dish of clean water is available to them. Do not use a deep dish, as the chicks may not be able to reach it and in their excitement to drink may fall in and drown.

At 6–8 weeks of age

When the chicks reach this age, slowly change their diet to 'grower's crumbs'. For a short period, combine the starter crumbs with the grower's crumbs; do not change suddenly from one food to another as this could be detrimental to young chicks.

At 16 weeks

The chicks can now be fed on an adult, egg-laying mixture, formed of either pellets or mash (see page 35).

Suitable temperature

Give the young chicks the same cosy and warm conditions provided for the hen when she was broody. Similarly, ensure that the area is dry, free from draughts and secure against vermin, especially rats.

Continue to provide the mother hen with warm and dry bedding, such as straw, hay or wood shavings. The chicks, especially when they are young, will snuggle up to the mother hen in order to keep warm.

Feathered up

At about 6–8 weeks, the young chicks will have developed their feathers, become established and reached a stage when they are said to be 'hardened off'. If the weather is mild, they can be put outside in a wire-netting run, complete with a protective part that provides shelter from rain and wind. This is sometimes known as a 'haybox brooder' (illustrated on page 59). A run about 6 ft (1.8 m) long, 60 cm (2 ft) wide and similarly high will accommodate up to 12 chicks.

Artificial incubation

Rather than relying on a mother hen to hatch fertile eggs (see pages 50–53), you can use artificial incubation to do the job. For this, you will need either a surrogate hen or an incubator to provide fertile eggs with the necessary temperature to encourage them to hatch and produce young chicks. Broody hens are not always available, especially early in the year, and therefore an incubator is often essential. It is easily installed and very efficient.

WHAT HAPPENS IF I DO NOT HAVE A COCK BIRD?

If you do not have a cock bird in with your hens, you can buy fertile eggs from a supplier and put them underneath a broody hen or in an incubator, which can be either bought or hired.

ADVANTAGE OF USING A BROODY HEN

The advantage of using a broody hen – and therefore going down the traditional route of producing chicks – is that it is natural, with the hen undertaking the role of a mother.

The advantages and disadvantages of using an incubator are described below.

INSTALLING AN INCUBATOR

Buy the best-quality incubator you can afford and have it installed by a qualified electrician. Place it in an even and constant temperature, away from positions where it will be knocked and jarred – also out of direct and strong sunlight that might cause it to overheat and damage the eggs.

Do not have electrical cables strewn around the room; if they are pulled or tripped over and the incubator damaged, you might lose the results of several weeks' work.

Eggs at various stages of hatching (including hatched chicks) will be apparent in incubators, especially large domestic ones.

ADVANTAGES AND DISADVANTAGES OF ARTIFICIAL INCUBATION

Advantages

- Enables eggs to be hatched even when a broody hen is not available.

- Once you have an incubator, you can easily raise fresh pullets to replace hens that are not producing the numbers of eggs expected of them.

- The cost of buying an incubator should be considered as a long-term investment. Sometimes they can be bought second-hand, but ensure that the electrics are safe and suited to your electricity supply.

Disadvantages

- The cost of buying an incubator can be high if you are only going to use it a few times. It might be less expensive to buy point-of-lay pullets that are just about to start laying eggs.

- When an incubator is bought and installed, it incurs cost and vigilance while the eggs are hatching. It is essential to maintain the desired temperatures (see below) during this period, as well as marking the eggs and regularly turning them (see below).

- You will need to install it in a vermin-proof, quiet and draught-free shed which has an electrical power supply (although incubators during earlier times were warmed by oil lights).

- You will need to carefully monitor the temperature (see page 55) for about three weeks to enable the chicks to hatch safely.

- The eggs will need regular turning (see page 57) – something that a mother hen does naturally. This needs to be done at least three times a day to prevent the embryo chick sticking to the inside of the shell.

TYPES OF INCUBATOR

There are many incubators available – some for large poultry establishments and others well suited to the needs of a home poultry enthusiast. The two main types are:

Small domestic three-egg incubator with microchip control

Still-air incubators

These do not have fans and therefore the air tends to stratify over the eggs. Unfortunately, the temperature is difficult to monitor or control in this type of incubator and it is very easy to overheat the eggs accidentally.

   The best way to assess the temperature – and therefore raise or lower it – is by using two thermometers and taking an average reading. The bulbs of the thermometers should be level or slightly above – up to 5 cm (2 in) – higher than the tops of the eggs) and not close to the source of heat.

Forced-air incubators

These have fans built into them that circulate the air at the correct temperature. The temperature should be set at 37.2–37.5°C (99–99.5°F). This will allow the temperature to rise slightly to 37.7°C (100°F).

Small domestic incubator with full temperature and humidity control

IF THERE IS A POWER SUPPLY FAILURE

Keep the eggs as warm as possible by placing several blankets of top of the incubator. Candles can be put in jars and positioned close to the incubator, but take care not to place flammable material near them. If the power cut lasts for more than a few hours, later 'candle' (see page 57) the eggs to check if they are still fertile.

THE CHOICE OF FERTILE EGGS FOR HATCHING

• Select clean, good-sized eggs from hens rather than pullets that have only just started to lay eggs.
• Preferably, the eggs should be about 70 g (2 oz) in weight.
• The fresher the egg, the greater the chance of hatching a chick. Therefore, they should not be more than seven days old (see below for storing eggs before incubation).
• Do not use extra large eggs, as they could contain two yolks and be unlikely to produce chicks.
• Avoid cracked and misshapen eggs; even hairline cracks allow infection to enter the egg.

STORING EGGS BEFORE INCUBATION

If you have eggs waiting to be placed in an incubator – which perhaps has yet to be positioned and set up – it is possible to store fertile eggs for up to 14 days (but, preferably, for no more than 7 days). Temporarily position each egg, pointed end downwards, in egg trays, in a temperature of 12.7–15.5°C (55–60°F) and with a humidity reading of 70–75%. Turn them three times each day. Temperatures in excess of 23.8°C (75°F) and with a humidity of lower than 40% will quickly damage them. If the humidity is too low, place a shallow dish of water close by.

TEMPERATURE, HUMIDITY AND VENTILATION

In addition to regularly turning the eggs (see page 57), temperature and humidity must not be neglected.
Temperature: The optimum hatching temperature is 37.7°C (100°F) and ideally this is the temperature that needs to be achieved at egg height. The embryos in the eggs become damaged – often fatally – if the temperature rises above 39.5°C (103°F) or falls below 35.5°C (96°F) for several hours.
Humidity: The relative humidity in the incubator can be influenced by changing the size of the water pan, or by putting a small sponge in the water to increase the surface area for water evaporation. Preferably, the relative humidity should be about 60% for most of the hatching period, but rising to 65–70% during the last three days (hatching usually takes 20–21 days). A wet-bulb thermometer can be used to assess humidity; remember that a rise in humidity affects the temperature, so you should vigilantly check these readings.
Ventilation: The eggs need normal atmospheric air, so check that ventilation is adequate and holes and slots are not blocked, preventing its entry and escape.

SHOULD I WASH EGGS BEFORE PLACING THEM IN AN INCUBATOR?

Unless the eggs are extremely dirty, do not wash them. Washing destroys the surface coating on the shell and could enable germs to enter through the minute pores in the egg's shell. Instead, just lightly brush off any dried dirt.

   However, if an egg is heavy encrusted with dirt, use a slightly damp cloth, with the water fractionally warmer than the egg's temperature. This enables the egg to sweat the dirt out of the pores. Take care not to use water that is cooler than the egg.

PUTTING EGGS IN AN INCUBATOR

- Switch on the incubator several hours before it is required, to ensure that it has reached the correct temperature (see page 55) by the time you need it.

- Before putting the eggs in the incubator, take each one and mark an X on one side and a circle on the other with a marker pen. This will enable you to readily identify the eggs that you have turned (see page 57).

- At the same time, write the expected date of hatching on each egg (allow 20–21 days from when the egg is placed in the incubator). This will later help to direct your attention to eggs that are soon to hatch.

- Do not cram a small incubator with a lot of eggs.

- Ensure that the water tray (which will be at its base) is regularly topped up with clean water, so that the correct humidity is maintained (see page 55 for details of the desired humidity).

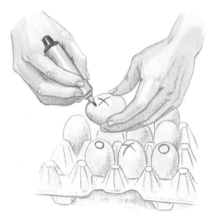

Above: Marking the eggs with indelible ink before you put them in the incubator will help you later. Put an X on one side and an O on the other, as well as the date of expected hatching.

Above: Maintaining the correct humidity is essential to ensure that the eggs are not dried by the warmth within an incubator (see page 55 for details).

CANDLING THE EGGS

Seven or eight days after putting eggs in an incubator, they need to be candled. This is an easy and quick check to see if an egg is fertile and therefore will produce a chick. If it is not fertile, the egg should be removed.

This task is known as candling because in earlier years eggs were tested by using a candle. Proprietary equipment is now available for doing this job, although it is possible to make your own 'candler'. This involves a box about 15 cm (6 in) square and 20–23 cm (8–9 in) long (see illustration on the right). Inside one end, fit a bulb socket and put a 60-watt bulb in place. At the other end, cut a 3 cm (1¼ in) wide hole.

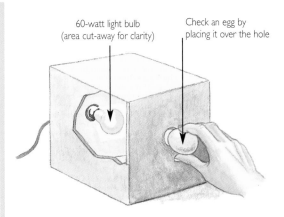

60-watt light bulb (area cut-away for clarity)

Check an egg by placing it over the hole

To candle an egg

• Turn out all lights in a dark room.

• Turn the light on in the candling box.

• Take the eggs, one by one, from the incubator and position over the hole from where the light is coming.

• Look directly down at the egg, or at a slight angle if this is easier.

• If the egg reveals a small, dark red spot with veins radiating out in all directions, it is fertile and can be returned to the incubator. Ensure that this testing is undertaken quickly to avoid cooling the egg.

• If the egg is clear, it is infertile. It can be boiled, chopped up and fed to your hens.

Home-made candling devices are easily constructed, but make sure that all electrical fittings are safe and properly fitted. For construction details, see above left. Do not cut the viewing hole at the top of the candling box, because the heat in the box will rise rapidly and may damage the egg.

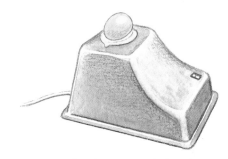

Proprietary candling devices are available and these are efficient and easily installed.

Candling after 7–8 days

Infertile
The egg shows up as perfectly clear and will not produce a young chick. Remove it and use to feed your hens.

Fertile
Small, spidery lines radiate from the centre. Replace it in the incubator as soon as possible in order to prevent it cooling.

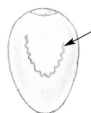

Damaged embryo
This sometimes occurs when the shell is damaged. It will not produce a chick and is best discarded.

TURNING EGGS AND HATCHING

The eggs will need regular turning – three or more times each day – to prevent the embryo chick sticking to the inside of the shell. The markings you have made on each egg will ensure that you know which eggs have been turned. Many poultry enthusiasts recommend turning an egg an odd number of turns each day; this ensures that an egg does not spend each night on the same side. You must stop turning the eggs on the 19th day – you will soon be able to hear the chicks using their beaks to peck at the inside of the shell. Each chick will continue pecking until it breaks free.

This is a time of great excitement in a household, and within a day or so chicks will appear. Many incubators are fitted with a panel through which you can see what is happening inside, so you can prepare for the great moment.

All chicks – whether those of guinea fowl (shown here) or chickens – need dedicated care when they hatch to ensure that they will develop into healthy birds.

CARING FOR NEWLY HATCHED CHICKS

What to do when chicks hatch and you have a hen to care for them is described on page 59. However, since incubator-raised chicks do not have a mother to look after them, they need special help; this is provided by an artificial brooder, which creates a cosy and warm home for them. They also need to be fed and watered regularly (see page 53).

Some home poultry enthusiasts like to leave the young chicks alone for 24 hours after they hatch to enable them to dry and become fluffy. They do not need to be fed during this short period as they will be relying on food eaten earlier in the egg.

CULLING THE CHICKS

It is essential to remove all weak, blind and deformed chicks; they will never develop into satisfactory hens. Killing young chicks is not a job for the squeamish, but it has to be done. One way is to sharply knock the chick's head. Another method is to break its neck against a piece of stout wood.

It is best to keep young children away from the area while you are despatching the chicks. The sight of baby chicks being killed could be deeply disturbing for them.

Artificial brooder

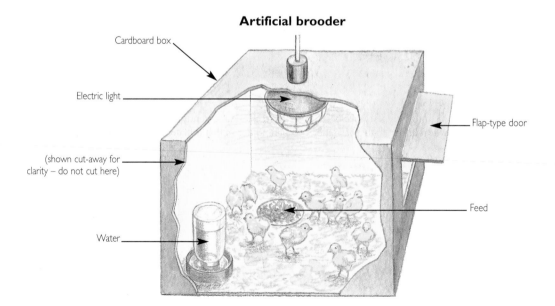

Cardboard box

Electric light

Flap-type door

(shown cut-away for clarity – do not cut here)

Feed

Water

Haybox brooder

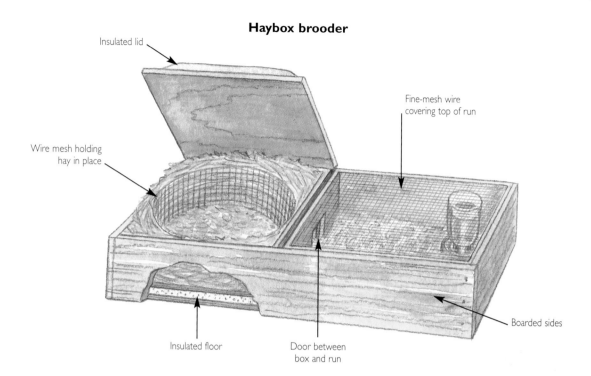

Insulated lid

Fine-mesh wire covering top of run

Wire mesh holding hay in place

Insulated floor

Door between box and run

Boarded sides

ARTIFICIAL BROODER

If the chicks cannot be found a home with a broody hen (see right), an artificial brooder is essential. This is a dry, warm, vermin-proof area – sometimes provided by a large, stout, cardboard box (see opposite page) – that restricts the chicks and provides a safe home for them.

To provide warmth, use an unshaded, tungsten-filament light bulb suspended 30–45 cm (12–18 in) above the brooder's floor. These bulbs are available in power from 100W to 250W; large areas require a large bulb to provide sufficient warmth.

Initially, the temperature in the artificial brooder needs to be about 35°C (95°F). When the chicks are established in their new home – usually after 2–3 days – the temperature should be slowly reduced by 1–2 degrees each week. Raising and lowering the light bulb will adjust the temperature at chick height. Alternatively, use a 'power reducer' to decrease the heat output of the bulb.

Carefully watch the chicks for signs of excessive warmth, indicated by them spreading to the edges of the box. Conversely, if they are cold they will cluster in the centre.

The chicks need to be introduced to food and water. If crumbs are scattered on the floor near the feeder, they soon get the idea of pecking and where the food is positioned. A suitable diet for chicks is described on pages 34–35. You may need to introduce the chicks to water by individually dipping each beak into it.

Continue lowering the temperature until the chicks are fully feathered, when they can be transferred into a 'haybox' brooder (see above).

BROODING YOUNG CHICKS WITH A BROODY HEN

If, by the time the young chicks hatch from an incubator, you have a broody hen available it is worth introducing the youngsters to her. However, she must have been broody for at least two weeks.

Some broody hens will accept an 'instant family', while others will not. If you are fortunate in finding a new mum, it will save you a great deal of trouble in having to transfer the chicks to an artificial brooder and raise them yourself.

The best surrogate mother for day-old chicks is a hen that is two or more years old. In the breeds section of this book (pages 10–27) an indication is given of breeds that make good mothers.

The introduction of youngsters to mum is best under-taken after dark, gently putting the chicks, one by one, behind her so that initially she is unable to see them. There is a chance that she will reject them by kicking backwards. If this happens, you will have to place them in an artificial brooder (see above and page 58) as soon as possible and before they become cold.

In most cases, however, she will accept the chicks and, if she is of a large breed, take up to 12 of them. An indication of her acceptance is when she clucks in response to their cheepings. She usually spreads herself over them, creating an instant family.

Leave her alone for a few hours, but you must regularly check on her to see if she has had second thoughts and rejected the chicks.

Keeping chickens healthy

Are there many problems?

There is a wide range of problems that can occur in chickens, from internal and external parasites to diseases and stress-related disorders. With all of them, keeping careful day-to-day observation of your poultry is essential, together with a rapid response when you notice that something may be wrong. Cleanliness and good feeding will help to prevent many of these troubles arising and reaching epidemic proportions in your flock.

SIGNS OF GOOD HEALTH IN CHICKENS

There are several indications that your chickens are in good health:

Behavioural signs
- Feeding and drinking normally
- Calm, but not silent
- Moving around normally
- Putting on weight – if young (usually up to the age of 18 weeks)
- Laying – if a hen and of the correct age
- Preening regularly
- Perching steadily and with confidence
- Sparring or mock fighting (in young birds)
- Sunbathing or dust bathing

Comb Correct colour and size

Eyes Bright and clear

Tail Carried correctly for the breed

Breathing Even, not rattling and without discharge

Feathers Shiny and smooth

Droppings Dark and firm and with a white tip

Gait Upright and alert

Smell It is correct if the bird smells warm and pleasant

Bodyweight Correct for breed and age of the bird

Legs Clean and strong

What are the types of problems?

This is a broad grouping of the range of problems – and where they are dealt with in this book.

Internal parasites (see pages 62–63)

- Caecal worms
- Coccidiosis
- Gape worms
- Gizzard worms
- Roundworms
- Tapeworms
- Thread worms

External parasites (see page 63)

- Fleas
- Lice
- Mites
- Ticks

Diseases (see pages 64–65)

- Avian Enterohepatitis
- Avian Influenza
- Fowl Cholera
- Marek's Disease
- Newcastle Disease (Fowl Pest)
- Roup

Problem habits (see pages 66–67)

- Bumble foot
- Cannibalism
- Cloacitis
- Cramp
- Crop impaction
- Egg binding
- Egg eating
- Feather pecking
- Internal laying
- Prolapse

Stress-related disorders (see page 65)

These encompass many symptoms, including diarrhoea, laboured and irregular breathing, and changes in normal activity rates. They all need immediate attention when noticed.

NOTIFIABLE DISEASES

These are diseases of chickens that must be notified to official government organizations as soon as they are noticed. At the same time, isolate the birds and keep people and animals away from the area. Notifiable poultry diseases in the British Isles include:

- Avian Influenza (Bird Flu) – see page 64
- Newcastle Disease (Fowl Pest) – see page 64

Please note: Many notifiable diseases are specific to certain countries. Therefore, in countries other than the British Isles check with your local animal health and welfare agency for details of them.

BODY TEMPERATURE

A chicken's normal temperature is 40–41.5°C (104–107°F). This temperature may fluctuate slightly in response to the environment. It does not usually go outside this band if the air temperature is 10 or so degrees less than the chicken's body temperature. However, during hot summers a chicken's temperature may rise dramatically. If it reaches the lethal zone of 45–47°C (113–117°F), the chicken must lose body heat quickly. Fortunately, chickens are well able to lose body heat. Their normal respiratory system, together with the ability to pant (in which they use the hot weather to evaporate water from their throats), enables their body temperature to be reduced fairly rapidly.

It is essential to ensure that the chicken house is well ventilated, and the roof insulated and able to reflect strong summer heat. There also needs to be shade in the run. If the temperature rises excessively, hens will stop laying and may even, eventually, die.

ASSESSING A CHICKEN'S TEMPERATURE

In earlier times, a chicken's body temperature was assessed by inserting a thermometer into its rectum. Nowadays, digital thermometers are available and make reading the temperature easier. Nevertheless, they still have to be inserted into a chicken's rectum.

Preventing the onset of pests and diseases

Good husbandry helps to prevent the onset of these problems. Here are some precautionary measures you can take:

- Keep their living areas clean – it is less expensive to clean chicken houses frequently than to suffer large veterinary bills and incur the loss of egg production.

- Make sure the birds are given the correct and balanced diet for their age (see pages 34–35 for details of feeding adult birds, and page 53 for chicks).

- Always store chicken food in a clean, dry and vermin-proof place – do not use food after its 'use by' date.

- Check daily that the drinking water is clean and not contaminated by faeces or decaying organic material.

- Although chickens like leftover kitchen scraps, do not feed them any that are mouldy or rotten.

- If you notice a chicken that appears to be ill, first isolate it in a separate enclosure. Then immediately seek veterinary advice.

- Do not cram a large number of chickens into a small area (see pages 32–33 for a range of chicken houses and arks, and indications of the amount of space they need).

WILD BIRDS

Veterinary experts suggest that some diseases can be transmitted from wild birds to domesticated chickens. For this reason, position feeders and drinkers within the confines of a hen house. If they are put in an open run, ensure that they have a raised lid so that faeces from wild birds cannot drop into them.

If you are anxious about wild birds having access to fresh water, provide a dish of easily accessible water specially for them – but away from your chickens.

Parasitic pests of chickens

What are parasitic pests?

These pernicious and tenacious creatures live either inside a chicken or on its outside, often causing extreme irritation, physical debilitation and loss of egg production. Treatment must be given immediately they are noticed; if ignored, they multiply rapidly and become even more difficult to eradicate. Indeed, repeated treatments are usually necessary to keep them under control. Vigilance is vital if you wish to keep your chickens healthy and productive.

INTERNAL PARASITES

CAECAL WORMS

Sometimes known as cecal worms, there are several species and they range in colour from yellow-white to white and are up to 18 mm (¾ in) long. They are found in the caeca (blind gut) of poultry and usually produce more difficulties for turkeys than for chickens. Indeed, this worm carries the organism that causes Blackhead disease (see Avian Enterohepatitis, on page 64) in turkeys. Chickens free-ranged with turkeys are especially at risk.
Symptoms: Usually not a problem with chickens; severe infestation can cause inflammation of the caeca, which results in pale birds with a droopy nature and huddling close together. They eat less food and drink less water, often accompanied by diarrhoea. They become emaciated and dehydrated, with a marked loss in egg production.
Treatment: Consult a vet and treat with a worming medicine.

COCCIDIOSIS

This problem is caused by several species of coccidia, which are small parasites that live in cells lining the gut, eventually destroying them. The problem is perpetuated and spread by the bird's droppings.
Symptoms: The problem develops into watery diarrhoea which may be stained with blood. Birds become ruffled-looking and distinctly sick. They also lose weight and become lethargic.
Treatment: Consult a vet, isolate the chicken and treat with a medicine. Free-range birds are more at risk. Good sanitation is essential, keeping the area clean and free from faeces. If the birds are in an ark, move it to fresh ground.

GAPE WORMS

Sometimes known as forked worms and red worms, they are 6–18 mm (¼–¾ in) long, red and Y-shaped. They infest many types of poultry, especially chickens, but also turkeys, pheasants and other game birds.
Symptoms: Birds gape, repeatedly opening their beaks without making any noise. They also shake their heads, have difficulty in breathing, are short of breath and display a loss of appetite and condition.
Treatment: Consult a vet, isolate the chicken and treat with a worming medicine. Birds that are kept as free-rangers are more at risk than those contained in a run. This is because worms, slugs and snails are hosts and able to reinfect poultry.

GIZZARD WORMS

Primarily a problem with geese, but it may infect chickens if they associate together. This parasitic worm uses beetles and grasshoppers as inter-mediate hosts, and free-range poultry in spring and early summer are most at risk.
Symptoms: Fine, reddish, thread-like worms, 12–25 mm (½–1 in) long, infect birds, causing them to stagger, lose weight and become weak. Older poultry tend to survive an attack, but young ones are more susceptible and need immediate attention.
Treatment: Early treatment is essential if the birds are to be saved. Consult a vet and add a worm medicine to the drinking water. Additionally, where possible keep poultry off contaminated land and especially where there is a communal pond.

ROUNDWORMS

Distinctive, yellowish-white, thick, oval and up to 10 cm (4 in) long worms that can be seen in the droppings of infected poultry. Most poultry can sustain slight infections of these worms, but it is when they build up to a high level that they cause serious problems.
Symptoms: Loss of condition, listless-ness, poor growth and diarrhoea. Young birds if not treated may just waste away.
Treatment: Consult your vet and either give pills to your chickens or add a worming medicine to their drinking water. Keep all feeding and drinking equipment clean and free from faeces. Also keep chickens off contaminated land; grasshoppers and earthworms often transmit roundworms.

TAPEWORMS

There are several different species of these bowel parasites. They are often flat, and segmented – each part is a developing egg case that when mature can break off and pass out of the host. This enables them to contaminate land and spread. Some tapeworms are up to 25 cm (10 in) long, and interposed in their life cycles are slugs, snails, beetles and flies. Should a roundworm-free chicken peck at and eat one of these, it soon becomes infected.
Symptoms: Rapid loss of weight, lethargy and difficulty in breathing.
Treatment: Roundworms are difficult to eradicate once established in a chicken's gut. Consult with a vet and introduce preventive treatments to both clear and prevent infections. Free-range birds are more likely to become infected that those solely kept in runs.

THREAD WORMS

These minute worms are sometimes known as capillary worms; several species affect different parts of a bird and result in many and specific symptoms. Mostly, they infect the upper part of the intestine. The small eggs of thread worms are extremely difficult to identify in the faeces.

Symptoms: General debilitation and diarrhoea, especially if the infestation is severe. Indeed, chronic infestations can ultimately cause death.

Treatment: Consult your vet as soon as possible and initiate treatment to clear the problem.

Local advice

For many home poultry keepers, a local chicken club or farm store is the first place to call for advice about worming medicines; but, if you are uncertain, always contact a vet.

EXTERNAL PARASITES

Fleas

Widely known and part of a large group of insects that infest mammals, from dogs and cats to chickens and the human body. They are small, they jump, and they have mouthparts that pierce and penetrate the skin, sucking blood. They transmit diseases.

Symptoms: The sucking of blood causes restlessness and itching. Fleas are often found in groups on the skin and, for protection, hidden away.

Treatment: As soon as they are seen (especially look for them if your chickens peck at their skin) dust them with a flea powder. Remove the chickens from their enclosure, take away all straw and bedding, and thoroughly scrub and wash the building, including nesting boxes. Spray the enclosure with an insecticide, then replace with fresh bedding. Inspect your chickens every few weeks to check if the fleas have become re-established.

Lice

Poultry lice – usually 3 mm (⅛ in) long and ranging in colour from off-white to brown and dark grey – feed by chewing dried skin and feathers, causing irritation and discomfort to the chicken. There are several types of lice that infest chickens, some on their body and others on the head.

Symptoms: Infected chickens are usually poor layers and become especially susceptible to diseases. The wings of the birds sag and droop, and they may become drowsy.

Treatment: As with the treatment for mites and fleas, remove the chickens from their enclose, spray or dust them with a recommended powder, and thoroughly clean the chicken house.

Ticks

They hide in cracks and crevices in a chicken house during the day and invade poultry at night. Ticks are small, tough and resilient.

Symptoms: They attach themselves to bare skin under wings or around the head, sucking blood and causing anaemia and a loss of egg production. It is very debilitating for poultry to be invaded in this way. Incidentally, the larval stages of ticks sometimes remain attached to birds for a long period.

Treatment: Remove all poultry from the enclosure and wash with a power-jet. Allow to dry, and then dust with an insecticide. Dust the poultry with an insecticide.

Suitable insecticides

Always check that the insecticide you are using is recommended for use with poultry. If you are not sure about its suitability, ask a vet for advice.

Mites

These are even more pernicious than fleas and certainly more persistent than lice. Like fleas, they feed on the blood of chickens, usually at night, and hide either on chickens or in the chicken enclosure during daytime, laying their eggs which, after about 10 days, turn into adults. Notches and crannies are an especial attraction for them; they can live for several months in a chicken shed, even when chickens have been removed. These mites resemble minute, near-invisible spiders. There are several types that attack chickens – the Red Mite or Roost Mite, and the Northern Fowl Mite.

Symptoms: They cause chickens a great deal of irritation and, in extreme infestations, can result in anaemia and death.

Treatment: Remove the chickens from their enclosure and treat with a proprietary chemical control recommended for use on chickens. Then remove and burn all bedding and thoroughly clean the enclosure. Water from high-pressure jets will dislodge them; when dry, spray or dust the enclosure with a insecticide.

Diseases of chickens

Are there many diseases?

There are several diseases that could be a problem for your chickens and two are notifiable (see below and page 61). In addition to those discussed here, there are many others, and their initial identification can be confusing and difficult as several diseases may appear similar in their early stages. It is therefore imperative to contact a vet if you are in doubt about any possible problem with your chickens and their continuing good health.

AVIAN ENTEROHEPATITIS

Also known as Blackhead and most often seen in turkeys, although occasionally in chickens (quails, guinea fowls, grouse and partridges are also affected). It is caused by a protozoan organism transmitted either in the droppings of an infected bird or in the eggs of a caecal worm (see page 62) and then excreted. These eggs survive in the soil for several months and are then often ingested by a bird.

Symptoms: It is called Blackhead because birds sometimes have a dark and discoloured head. Other symptoms include drowsiness, drooping wings, stilted gait, carrying the head down, and closed eyes.

Treatment: It is usually fatal in turkeys up to the age of 12 weeks. You should always keep turkeys on ground separate from chickens to reduce the risk of reinfection. Also, regularly change the litter and use anthelmentic drugs in the water or feed. You may need to consult a vet.

AVIAN INFLUENZA

Widely known as Bird Flu, it is a notifiable disease (see page 61); chickens and turkeys are especially at risk, although all poultry can become infected. It can be spread by wild birds.

Symptoms: These vary and depend on the age and species of poultry, but the usual signs are depression and droopiness, sudden decrease in egg laying, loss of appetite, diarrhoea, blood-tinged discharge from nostrils, distress when breathing, swelling of head, eyelids, comb, wattles and hocks, and lack of coordination when walking and standing.

Treatment: At the first sign of attack, immediately contact your local Animal Health office. Isolate the chickens and ensure that – until you are given permission – no one leaves or enters the premises.

FOWL CHOLERA

Disease of chickens, as well as turkeys, ducks and geese. It is highly contagious and unless treatment is given quickly can be fatal. Indeed, a hen may appear perfectly healthy one day, but be found dead the next.

Symptoms: Greenish-white diarrhoea, ruffled feathers, coughing, severe thirst, swollen joints, lameness, loss of appetite and a nasal discharge are just some of the symptoms.

Treatment: Isolate infected chickens immediately and contact a vet as infected birds must be killed and disposed of in a legal manner; do not just bury them. There are preventative drugs and these are added to the drinking water.

MAREK'S DISEASE

Also known as Fowl Paralysis Virus, this is highly contagious and usually results in the death of chickens unless they have been vaccinated when day-old chicks; immunity develops within two weeks.

Symptoms: Birds are usually 4–8 weeks old at the onset of symptoms, which include paralysis of the wings, neck and legs, and loss of weight. Birds usually have a characteristic 'straddling' stance, with one leg held forward and the other back, and with progressive paralysis. Vision is also impaired, showing as grey irises and irregular pupils.

Treatment: This is not possible after onset of the disease. Vaccinate young chicks (see previous column). There is currently work being carried out on the introduction of resistant strains – but this is for the future.

NEWCASTLE DISEASE

Sometimes known as Fowl Pest, in many countries this is a notifiable disease (see page 61). However, the term Fowl Pest can be confusing as it also refers to Fowl Plague. Therefore, for clarity it is best to refer to this disease as Newcastle Disease. It is a major killer, especially of young birds in large, commercial flocks; it can affect chickens suddenly and without warning. It also attacks turkeys.

Symptoms: Sneezing, gasping for air, listlessness and drop in egg production are a few of the symptoms. You can also expect green diarrhoea and misshapen or soft-shelled eggs, together with nervous disorders such as twitching or spasms.

Treatment: There is no cure, nor treatment once it has attacked. Immediately contact your local Animal Health office and do not move the birds. Isolate the chickens, ensuring that – until you are given permission – no one enters or leaves the premises.

ROUP

If a common cold is neglected it may turn into Roup, which can be seen in two forms – 'Wet Roup' and 'Dry Roup'. Both forms are contagious. Roup is also associated with Vitamin A deficiency and sometimes known as Nutritional Roup.

Symptoms: Wet Roup results in an offensive discharge from a chicken's respiratory tubes. Dry Roup is similar, but with comparatively little discharge. Your chickens will also experience poor growth and feathering, drowsiness and eyelids that are often stuck shut by a thick exudation.

Treatment: Add Vitamin A to the drinking water. If possible, keep infected chickens isolated. In severe cases, consult with a vet. The disease is a distressing one for the chickens, as well as to observe.

STRESS-RELATED DISORDERS IN CHICKENS

Chickens delight in a restful, quiet and unchanging routine, where food arrives at the usual times each day and is provided by the same person. Poultry put under stress are more likely to catch a disease and to show a decrease in egg laying.

Signs of stress
- Diarrhoea
- Laboured and irregular breathing
- Changes in normal activity rates

Preventative measures
- Handle chickens gently but with authority.
- Never put an excessive number of chickens in a hen house (see page 32 for the recommended density).
- Ensure they are correctly fed and that there are sufficient feeders and watering points for them all to feed and drink at the same time. This eliminates rivalry between them.
- Avoid loud noises – some poultry enthusiasts are said to play restful and soothing music to their chickens, but quietly!
- Do not move quickly and suddenly when among them.
- Ensure that the temperature is not excessively high and, if necessary, erect a hessian shade in a corner of the run. See page 61 for details of what a chicken's normal temperature is.

Above: Chickens normally have a friendly nature and soon learn to live in harmony with domestic pets – as well as bees.
Below: Chickens lay the most eggs when in a quiet situation and with a regular and unchanging routine.

Problem habits in chickens

These are problems that are not normally initiated by pests and diseases. Instead, they are difficulties that chickens occasionally encounter – such as crop impaction, egg binding and feather pecking – and usually they can be home treated. However, if you are puzzled about a problem and cannot readily identify it, contact a vet at the earliest possible moment. A keen and vigilant eye will save you a lot of money and reduce stress for your chickens.

Bumble foot

The bird hobbles and limps, following the other birds which are usually able to walk properly. As the name suggests, a bird with this problem appears to have a bumbling gait.

Symptoms: Usually the result of a bird's foot becoming cut, with the wound healing on the outside but infected internally. It results in a hard abscess on the ball of the foot that is painful, preventing the bird walking properly.

Treatment: Cut open the infected area with a sharp, sterilized scalpel and squeeze out the pus; ensure that the central, hard core is removed. When it is free of the infection, apply an antiseptic. Place the bird in a clean, dry pen until the wound has healed, usually after 3–4 days.

High perches sometimes initiate bumble foot, the bird landing awkwardly on a hard floor (see page 33 for the recommended height and spacing of perches).

Cannibalism

This is when a bird pecks, tears and consumes the skin and tissue of other birds in its group. It can occur in all types of housing systems, including free-range flocks. It occurs in ducks, turkeys, quails and pheasants as well as chickens. Primarily it is caused when birds are overcrowded or do not have sufficient food. It may also happen if the pecking order is not established, or when a new bird is introduced.

Symptoms: It often begins with feather pecking (see page 67), usually on the body, then spreading to the toes, tail and rear of a chicken.

Treatment: It is much easier to prevent cannibalism than to control it once established. Always give birds sufficient space (see page 32 for the area each bird needs), as well as enough food. Also ensure that there are sufficient feeding spaces for all birds to feed at the same time. A change in diet to a less tasty food may also trigger cannabalism.

Underweight birds are sometimes at risk; if this happens, move vulnerable birds to another pen.

Cloacitis

Sometimes known as 'Vent Gleet' and 'Pasted Vent', this is the discharge of a milky-white substance from the vent area at the rear end of a chicken. Although often thought to be a general problem, it is an infectious and highly contagious disease. Isolate affected birds immediately and, preferably, call a vet. Usually it results in the chicken being killed. Do not breed chicks from eggs produced by infected hens.

Internal laying

This is much more serious than egg binding (see page 67); it is when an egg takes the wrong channel within a hen, becoming infected and leading to peritonitis. Little can be done for the hen and she is best put out of her misery.

Cramp

Common problem, but one that affects ducks more than chickens. It arises through imperfect blood circulation and is aggravated by damp and cold conditions underfoot.

Symptoms: The bird uncontrollably squats down, usually suddenly, with its toes contracted.

Treatment: Move to drier conditions, so that feet do not remain wet. A layer of long straw reduces the possibility of dampness, at the same time keeping the bird's feet warm. Where possible, isolate an affected bird, so that it is quiet, dry and warm.

The problem is intensified by deficiencies of calcium or vitamin D in its food; add supplements to the bird's diet.

Crop impaction

Sometimes known as 'crop bound', this occurs when there is congestion in a bird's crop; it becomes packed with food that cannot easily be passed. It is often caused by pieces of long grass that have become twisted into a ball – keep grass short throughout summer if birds are free-ranging. Occasionally, a bird chews a piece of string, so never leave pieces where chickens forage.

Symptoms: The crop can be seen to be distended with fermenting food;, and the bird will not be eating in its normal way.

Treatment: Give the chicken a drink of warm water; this will distend the crop which then can be rubbed to ease the blockage. Alternatively, hold the hen upside down and gently squeeze the crop to remove the messy blockage. Then, allow her to drink freely but do not give her food for 24 hours.

Egg binding

This laying disorder arises because the oviduct (the long tube where eggs form and which delivers them to the hen's rear, ready for laying) is too small. This especially applies to pullets when first laying eggs. The other cause is when an egg has become broken within a hen and will not slip out easily.

Symptoms: A hen cannot lay an egg easily and properly.

Treatment: Carefully hold the hen for a short time with her egg-laying vent exposed to steam from boiling water, but taking great care not to harm her. The steam eases and softens the pelvic bones. Also give the hen olive oil and place her in a straw-lined box in a quiet area. After a few hours, she usually manages to expel the egg.

Incidentally, hens when first laying eggs may initially produce rather long and narrow ones, possibly streaked with blood. This problem usually disappears within a week or two.

Feather pecking

A problem usually associated with hens kept in runs, rather than free-rangers. Overcrowding and a poor diet are the main causes of this problem. It may also begin through sheer idleness and perverseness on the parts of chickens.

Symptoms: Areas mainly pecked are the rump, back and tail. The problem can become serious and lead to cannibalism (see page 66), especially if flesh becomes torn and bleeds.

Treatment: Remove seriously affected birds and place in isolation. Hanging up green stuff and giving extra feed to a group of chickens helps to allay the problem. If the problem continues, birds may need to have their beaks trimmed.

Prolapse

Sometimes known as 'down behind'. It does not happen to young hens, but after their second year of laying eggs a portion of the oviduct may protrude. It is caused by a hen laying oversized eggs, or through excessive straining to lay eggs. Overweight hens are especially likely to incur this problem.

Symptoms: Part of a hen's egg-laying organs protrude out of her backside.

Treatment: This is a problem that needs to be treated quickly as it may lead to cannibalism (see page 66) in the rest of the group. The comfort of the bird also needs to be restored.

Support the hen, with her head downwards, and use warm water to gently wash and clean the area. Then apply some olive oil, using a finger to work it into the orifice. Place the hen by herself in a warm box for about a week – in a quiet and lightly shaded area. Do not excessively feed the hen during this period, but give plenty of water. If the problem persists, repeat the treatment – several times if necessary.

Egg eating

This problem usually begins as an accident, when a hen lays an egg and it breaks open on hitting the floor. Hens are naturally inquisitive and bustle around the egg to taste it, which they usually find attractive. They then start to break their own freshly laid eggs. The problem can also begin when a hen lays an egg without a shell. Nest boxes that are too bright and in direct sunlight sometimes initiate this problem.

Symptoms: It usually only happens by chance when an egg breaks, but occasionally congested hens start to eat eggs, especially if they are bored.

Treatment: Give your hens plenty of space and something to occupy them (to relieve boredom); hang green stuff around the run to attract their attention. A lack of calcium in their diet can make older hens peck at eggs and eat them, so add supplements to their diet if necessary.

Animal pests

- **Cats:** Family cats usually live in harmony with your chickens, but aggressive local cats will need to be watched.

- **Dogs:** Once your chickens are established as 'new boys on the block', most dogs take no notice of them.

- **Foxes:** A continuing threat and only soundly constructed fences will keep them out.

- **Mice:** More likely to threaten stored food than your chickens.

- **Rats:** Pernicious, intrusive vermin, attacking chicks and chickens. Regularly security-check all enclosures.

Keeping ducks

Ducks are less demanding and easier to keep than chickens, as well as being less susceptible to some poultry diseases. A few breeds, such as Khaki Campbell, produce more and larger eggs than chickens – up to 300 eggs a year is not uncommon. Ducks' conversion of food to bodyweight is much higher than in most chickens; in ten weeks an Aylesbury duck can weigh at least 2.3 kg (5 lb). Therefore they should be considered as an alternative to chickens.

Ducks are full of self-importance and character, and will never fail to keep you amused; they also lay superb eggs.

GETTING THE TERMS RIGHT

Females are known as hens or ducks, while males are drakes.

WHAT ARE DOMESTIC DUCKS?

These are breeds kept for their abilities to lay eggs, produce meat for eating, or both. There are many breeds to choose from and several are featured on the opposite page.

WHAT ARE ORNAMENTAL DUCKS?

These are breeds of duck kept for their ornamental qualities; they include Mandarin, Falcated Teal, Garganey Teal, Pintail and Common Widgeon. There are many others and they are all attractive, especially when seen swimming on a pond. Like domestic ducks, they are best kept in rural situations, where they have good access to water as well as areas in which to wander.

HOW LONG DO DUCKS LIVE?

The average life span of a duck is about five years, although some live for up to ten. Their egg-laying slowly diminishes and invariably there comes a time when it is not profitable to keep them. Indeed, egg production sometimes falls by 25 per cent in their second year of laying, with a similar drop during the following year.

ARE DUCKS SUITABLE FOR SMALL YARDS AND GARDENS?

Ducks are not suitable for small gardens – especially in built-up areas – as their quacking will irritate neighbours. They are also messy eaters, and when in small runs they can create a quagmire during wet weather. However, if you have a large garden in the country – especially where they can either wander through a field or be in a large run in a coarsely grassed area – they are ideal. If allowed to wander in cultivated areas, their webbed feet will soon break down and destroy plants.

CARRYING A DUCK

It is essential to hold a duck firmly, so that it feels secure. Cradle it over an arm and securely hold its legs, but ensuring they are not harmed. A hand over its body helps to reassure the duck that it is safe and will not be dropped. Always put a duck down gently, so that its feet are firmly on the ground before you release them. Do not just drop a duck.

BREEDS OF DOMESTIC DUCKS

The range of domestic ducks is not as wide as that of chickens. Here are six of the most popular breeds:

Aylesbury

Glossy, snowy-white breed that originated in the Vale of Aylesbury, England. It attains a large size and weight at an early age, with good-quality and well-flavoured flesh.

Eggs: Bluish-green; 80–100 each year.

Weight: Hen 4 kg (9 lb), drake up to 4.5 kg (10 lb).

Indian Runner

Its upright stance makes it easy to identify, and until the Khaki Campbell was introduced it was one of the most popular breeds. There are several colour variations: Black, White, Fawn and White, Fawn, and Chocolate. It originated in India and was introduced into Britain about 100 years ago. Very hardy.

Eggs: White; about 180 each year.

Weight: Hen 1.8 kg (4 lb), drake 1.8 kg (4 lb).

Khaki Campbell

Originated in 1901 from a cross between an Indian Runner and a wild Mallard duck; later, Rouen blood was introduced into it. Although a prolific layer, the hens do not make good mothers and their meat is said to be stringy. Mature drakes are khaki, with a lighter colour underneath.

Eggs: White; about 300 each year.

Weight: Hen 2 kg (4 lb): drake 2.3 kg (5 lb).

Orpington

Dual-purpose breed that originated from the Indian Runner and Aylesbury. It is large and heavy, buff-coloured and ideal to keep where space is limited. Their ducklings are large enough to eat at 8–10 weeks. However, it is a breed that is not now so popular.

Eggs: White; 230–240 each year.

Weight: Hen 3.2 kg (7 lb), drake 3.2 kg (7 lb).

Pekin

Imported into Britain in the 1870s, it is an excellent table breed as well as a layer of eggs. It is sometimes crossed with an Aylesbury drake to produce an excellent dual-purpose breed. The Pekin is hardy and a superb forager of food. It is cream, with a bright orange bill, shanks and feet.

Eggs: Blue; about 120 each year.

Weight: Hen 3.6 kg (8 lb), drake 4 kg (9 lb).

Rouen

Hardy breed that originated in France and is ideal as a table bird. It is descended from the Mallard and has the same colouring; in summer, the drake's bottle-green head and brownish-red markings are lost in its moult. Because of this breed's decorative qualities it is often seen among groups of ornamental ducks.

Eggs: Greenish-blue; about 90 each year.

Weight: Hen 4.5 kg (10 lb), drake 5.4 kg (12 lb).

Housing and looking after ducks

Do ducks need special sheds?

Whether they have been store-bought, home-made or adapted from existing sheds, duck houses must be draught-proof, dry, well-ventilated and vermin-proof. Ducks are generally considered to be hardy and able to survive almost entirely in the open air, but there is no doubt that the best results are obtained by housing them correctly. Some of the most efficient duck-keeping systems are recommended below.

Ducks are useful lawn mowers and three are more than enough to keep 0.4 hectares (1 acre) of grass short.

HOW MUCH SPACE DOES EACH DUCK NEED?

A duck house needs to provide each duck with 0.37–0.46 m^2 (4–5 sq ft) of floor space. Unlike chickens, ducks do not need perches; instead, dry bedding is essential and this can be chopped straw, dry peat or sawdust.

SYSTEMS FOR KEEPING DUCKS

Several ways to keep ducks have been evolved. Some are suited to large areas, others are for commercial purposes, and some suit small areas and gardens. These systems are:
- **Free-range:** Large in scale and needing 0.4 ha (1 acre); this accommodates up to 100 ducks.
- **Dutch system:** Commercial, with duck pens and either a stream or channelled water running along one side.
- **Straw yards:** Commercial, with access to a regular and plentiful supply of straw bales.
- **Sun balcony system:** Also called the Veranda method, with ducks mainly kept in a suspended veranda formed of galvanized wire mesh. To many duck enthusiasts it is unnatural, with overtones of battery hens.
- **Small duck house:** Most often used by home enthusiasts and usually a modified shed or a bought one (see below).

A SMALL DUCK HOUSE

Whether modified from an old shed or custom-built, there are several essential features about the construction of a small duck house:

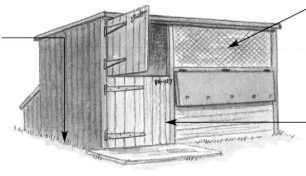

Flooring: Concrete provides a vermin-proof base, but can be cold, especially in winter. Rammed earth is a possibility but add a wire-netting sub-floor to prevent rats burrowing into the building. A wooden, slatted floor – raised about 5 cm (2 in) above the ground – is often recommended, although it makes cleaning more difficult. Again, it needs a defence against rats.

Ventilation: Fresh air circulating throughout the house is essential – but not a continual draught. A wire-mesh screen secured at the top of the house will enable air to enter and escape.

Doors and windows: Strongly latched doors are essential to prevent the ducks barging out. Hinged window openings must be vermin-proof.

STRAW NEST BOXES

These are essential and should be positioned on the floor of the duck house; secure them in place by using bricks or thick pieces of wood.

LOOKING AFTER DUCKS

Every morning:
- Collect the eggs.
- Loosen up and rake over the litter; as soon as it becomes wet and foul, change it. Ducks are prone to rheumatism and will be reluctant to enter a dirty, wet shed to sleep. Usually, completely removing the bedding and cleaning the shed every week is essential. However, some duck-keeping enthusiasts use a 'deep-litter' system, which means topping up the litter every few days and replacing it completely after about six months.
- Refill water containers with clean, fresh water.
- Ensure that grit is available.

CAN I USE DUCK MANURE ON MY GARDEN?

Yes, but not immediately. There are strong chemicals in it and it is best to add it to a compost heap first and then later dig it into the soil when it has decomposed and is part of other organic materials.

FEEDING DUCKS

- They require about 25 per cent more food than chickens.
- Each duck needs 200–225 g (7–8 oz) of food every day, half in the morning and half in the early evening when they are 'put to bed'. If your ducks are not laying properly, slightly increase the amount of food given to them.
- Free-ranging ducks will find food in the grass and soil, including slugs and insects, but ducks need more food than this if they are to produce plenty of eggs.
- Up to three-quarters of their daily food input can be cooked vegetables, including potatoes, swedes and carrots. Mix this with water and dry mash to form a sloppy but slightly firm mixture – it should not be runny.

IS IT NECESSARY TO HAVE A POND?

If you are keeping ducks for their eggs as well as fattening them for eating, a pond in which they can swim is not necessary. However, should you wish to breed ducks, a pond is essential in order to ensure successful mating. This applies especially to the heavier breeds, which need the buoyancy created by water to help a drake (male) mount the hen (female).

Even if a pond is not needed, a deep trough full of water is essential to enable them to cover their heads in clean water. If they are unable to do this, they may develop eye and nose problems. They also appreciate being able to splash their feathers with water.

DO DUCKS NEED SHADE FROM STRONG SUNLIGHT?

Protection from strong and direct sunshine during hot summer days is essential and this is usually provided by trees and shrubs. At night – and until mid-morning – your ducks will be in a duck house. After that, and depending on how they are kept, they can be let out to wander among the bushes and trees.

ARE STRONG WINDS DAMAGING TO DUCKS?

Yes. Ducks detest strong and cold winds, which can decrease egg-laying by a third. For this reason, position your duck house to face towards the sun, with a small, wire-netting run attached on the warm side. Ducks should be kept penned up overnight and until about 10 am to make sure that the eggs are laid within the duck house rather than among shrubs and bushes, where they are usually difficult to find.

DO I NEED TO CLIP THEIR WINGS

Where ducks are kept in pens, it is essential to clip their wings. However, remember that very large breeds such as the Aylesbury are usually too heavy for flight.

Wing clipping is best undertaken during the first ten days of a duck's life, and is achieved by cutting off the last 12 mm (½ in) from the tip of one wing only. With older birds, the top 6 cm (2½ in) from the flight feathers on one wing can be cut off.

There is a difference of opinion among duck enthusiasts about the amount of wing feathers that can be safely cut off. Therefore, before wing clipping your ducks ask an experienced duck keeper for advice about tackling this task.

Few sights are more pleasing than ducks splattering about in a pond or large tub of water, when they appear excited and totally engrossed in their aquatics. It reinforces the old adage 'just like a duck to water'. A pond is not essential (see above), but some form of access to clean water is.

Keeping quails

What are quails?

Quails are part of the pheasant family and are classified as game birds. Like pheasants, they are able to fly, and some are migratory. It is claimed that there are more than 100 breeds of quail in the wild, mostly found in North America and Asia. For this reason they can be divided into 'Old World' and 'New World' types. The majority of domesticated quails have been bred from Asian types and in particular from Japanese strains.

POPULAR VARIETIES

Japanese Quail

Bobwhite

The Japanese Quail (*Coturnix japonica*) is probably the most widespread and widely kept variety. It often lays 300 eggs in a year, and after six weeks of growth will be around 140–170 g (5–6 oz) in weight and ready for the oven. Hens (female quails) start to lay eggs at six weeks old.

The Bobwhite (*Colinus virginianus*) is widely kept in North America, especially for its meat. Other quails popular in North America include the large Mountain Quail (preferring brush clearings or forested areas), and the Gambel's Quail (also known as the Desert Quail).

Quails are best kept in rural areas full of bushes and trees, where they are able to pursue their natural inclinations.

KEEPING QUAILS

They are at their best when kept in a 'game' environment, with access to fields, bushes and trees. However, they require protection from wind and rain, and also need shade when the sun is strong. They are often kept intensively in cages, sometimes in suburban sheds and with 40 birds crammed into very small areas; to many environmentalists this smacks of battery-kept chickens, and is something to be avoided.

FEEDING AND WATERING QUAILS

Quails naturally feed on insects and seeds, nesting on the ground and seeking out hidden places where they will not be disturbed. However, when startled they immediately fly upwards, sometimes knocking into trees. Clipping their wings (see page 71) can solve this problem and restrict them to a smaller area.

Keeping guinea fowl

Native to Africa but widely domesticated, guinea fowl are seed-eating, ground-nesting, partridge-like birds, sometimes known as guinea hens. They have featherless heads, and are kept for their meat and their eggs. The birds can be noisy and especially troublesome to neighbours when disturbed and alarmed. Their piercing shrieks, in two tones from the hen (female) and a single one from the cock (male), have given them the reputation as good watchdogs.

What are guinea fowl?

GETTING THE TERMS RIGHT

An adult male guinea fowl is known as a cock, a female is a hen, and a youngster is a keet.

HOUSING GUINEA FOWL

Less adventuresome than quails and more easily enclosed, guinea fowl are sometimes kept in the same type of caged run as used for chickens. As such, they each need 18–20 cm (7–8 in) of roosting space and 0.2 m² (2 sq ft) of floor space.

MEAT COLOUR?

Most guinea fowl have dark meat, but the white varieties produce white meat.

This hen guinea fowl, with her large brood of keets (youngsters), enjoys searching the ground for seeds, a task that she undertakes in a purposeful manner.

POPULAR COLOURS

Buff Dundotte: Soft tan, with an even covering of white dots.

Chocolate: Beautiful and unusual dark brown, with a few dots and bars in the flank region.

Coral Blue: A medium blue, darkening to coral blue on the neck, breast and back.

Lavender: Light blue, with white dots.

Pearl Grey: Dark grey with white dots on the plumage.

Powder Blue: Uniformly light blue, with no barring or dotting from another colour.

Royal Purple: Very dark and black, with a purplish sheen.

Violet: Dusty black with a purple sheen.

White: Pure white with a few black hairs on the back of the neck.

Vegas White Guinea Fowl

Glossary

Addled egg
A fertile egg in which the embryo has died after starting to grow. Sometimes, the term just refers to any egg that has started to go bad.

Aerobic
Describes microbes that require oxygen to develop and multiply.

Air cell
Also known as air space and air sac. It is found in the large end of an egg and between the shell membranes, where it provides a chick with air prior to hatching.

Air sac
See Air cell.

Air space
See Air cell.

Albumen
The white part of an egg, as distinct from the yolk (yellow).

Alektorophobia
Fear of chickens.

American breeds
Breeds developed in America, with common characteristics such as yellow skin, non-feathered shanks and red earlobes.

American Standards of Perfection
A book published by the American Poultry Association describing each breed recognized by that organization.

Amino acids
The form in which proteins are absorbed into the bloodstream.

Anaerobic
Describes microbes not requiring oxygen to develop and multiply.

Araucana
A breed of chicken, known in North America as the South American Rumpless (see page 10 for details).

Ark
A small, portable chicken house with a ridged roof. It is suitable for a small number of birds and can be easily moved from one position to another.

Autosexing
Describes breeds in which the sex (male or female) can be determined in day-old chicks.

Axial feather
Short wing feather found between the primary and secondary feathers.

Baby chick
A newly hatched chick, before it has been fed and given water.

Balanced ration
A diet containing all the ingredients necessary for healthy growth.

Bantam
A diminutive chicken: some are bred down in size from standard-sized breeds, while others are naturally small fowl. True bantams do not have large counterparts. A range of true bantams is described and illustrated on pages 24–27, together with a list of standard-sized breeds that have bantam-sized counterparts (see page 23).

Barnevelder
Breed of chicken (see page 11 for details).

Barnyard chicken
A chicken of mixed parentage.

Barred Rock
Breed of chicken (see page 10 for details).

Barring
Alternate light and dark stripes across a feather. Some breeds of chickens, such as the Barred Plymouth Rock, have this distinctive characteristic.

Battery
An intensive system of keeping hens in which they are kept close together, usually in wire cages.

Beak
The hard, protruding part of a bird's mouth, formed of an upper beak and a lower beak.

Beard
Bunch of feathers under the throats of some breeds, such as Faverolles. See also Bib.

Bedding
The material – ranging from straw to wood shavings – used to form comfortable bedding for chickens and other poultry.

Belgian
Breed of bantam (see page 26 for details).

Bib
Another name for beard, or it may refer to white markings on specific breeds of duck.

Biddy
A laying hen that is over one year old.

Black Australorp
Breed of chicken (see page 11 for details).

Black Rock
Breed of chicken (see page 12 for details).

Black Silkie
Breed of chicken (see page 12 for details).

Blast freezing
Commercial method of freezing, when extremely cold air is blown over fowls in a chamber or tunnel.

Blood ring
Signifies early embryonic death.

Blood spot
Blood in the white part or yolk of an egg, caused by slight rupturing of blood vessels during its formation. It is detected by candling and is harmless to the eater.

Bluebelle
Breed of chicken (see page 12 for details).

Blue Laced Wyandotte
Breed of chicken (see page 13 for details)

Blue Silkie
Breed of chicken (see page 13 for details).

Boiler
A boiling hen – an old hen that requires thorough cooking to make it suitable for eating.

Broiler
A young bird (usually less than eight weeks old) specially raised to be killed for its meat.

Brooder
A heated piece of equipment for the artificial rearing of young birds.

Broody
A chicken with the natural urge to sit on eggs to hatch them.

Buff Cochin
Breed of chicken (see page 13 for details).

Buff Orpington
Breed of chicken (see page 14 for details).

Buff Sussex
Breed of chicken (see page 14 for details).

Bumble foot
A problem habit in chickens (see page 66 for details).

Caecal worms
Internal parasite of chickens (see page 62 for details).

Candling (or 'to candle')
A means of determining the interior quality of an egg by examining it against a strong light (originally a candle). See page 57 for details.

Cannibalism
The habit of killing or eating other birds in the flock. It can be the result of having too many birds in the flock, or through stress factors (see page 66).

Cap
A comb (see Comb); also the back part of a fowl's skull.

Cape
Feathers under and at the base of the neck hackle (see Hackle) between the shoulders.

Capon
A male chicken that has been feminized – either by surgery or chemically – to improve its speed of growth and eating quality (see page 38).

Checks
American term for cracked eggs.

Chick
A chicken from the time it hatches until it feathers out.

China eggs
Also known as 'pot eggs' or 'crock eggs', they are dummy eggs for placing under a hen to encourage her to lay eggs.

Clears
Incubated eggs which reveal no signs of embryonic development when candled (see Candling).

Cloacitis
A problem habit in chickens (see page 66 for details).

Clubbed down
An embryonic abnormality in which down (the soft, first feathering of a young bird) appears in small, beaded nodules.

Clutch
The number of eggs that a chicken, when in the wild, would lay before sitting on them. Nowadays, it usually refers to the number of eggs a poultry keeper allows a hen to sit on for hatching.

Coccidiosis
Internal parasite of chickens (see page 62 for details).

Cock
Male bird aged 12 months or more.

Cock bird
Male bird aged 12 months or more.

Cockerel
Male bird under 12 months of age.

Comb
The fleshy prominence on top of the head of a fowl, especially on the cock bird.

Concentrates
Additives that provide a ready-made protein, vitamin and mineral balance when mixed with a chicken's food.

Cramming
An early method of fattening birds by force feeding.

Cramp
A problem habit in chickens (see page 66 for details).

Cream Legbar
Breed of chicken (see page 14 for details).

Crest
A crown or tuft of feathers on the head of a chicken. It is sometimes known as the top-knot.

Croad Langshan
Breed of chicken (see page 15 for details).

Crock Eggs
See China eggs.

Crop
The enlargement in the gullet where food is stored and prepared for digestion (see page 34 for details).

Crop-bound
Describes the clogging-up of the crop through bits of grass or, occasionally, hay and straw.

Crop impaction
A problem habit in chickens (see page 66 for details).

Cross-bred
The first generation from crossing two different breeds or varieties. See also Hybrid vigour.

Cuckoo Maran
Breed of chicken (see page 15 for details).

Cull
To remove and kill infirm, ageing or surplus birds or chicks.

Cuticle
Bloom or residue that forms on a newly laid egg as it dries.

Dam
Occasionally used to describe a mother hen.

De-beaking
Cutting off part of the beak to prevent feather pecking and cannibalism.

Down
The soft, first feathering of a young bird.

DPK
Domestic poultry keeper.

DPL
Dried poultry litter, a mixture of manure and floor material from floor-housed birds.

DPM
Dried poultry manure.

Drake
Male duck.

Droppings
Chicken manure.

Dual-purpose breed
Breed of chicken that both lays good eggs and is excellent as a table bird (see page 8 for a list of suitable birds).

Duck
A female duck (a male is a drake).

Dust bath
A depression or box in the ground filled with dry earth, sand or sawdust in which hens can keep clean and cool.

Dusting
The act, on the part of a chicken, of thrashing around in the dirt to clean its feathers and to discourage body parasites.

Dutch
Breed of bantam (see page 26 for details).

Egg binding
The blockage of the oviduct by an extra large egg or an egg which has broken internally (see page 64 for details).

Egg eating
A habit of some hens of eating their own eggs (see page 67).

Egg tooth
A horny protrusion on a hatching bird's beak which helps it to break the shell when hatching. Later, it falls off.

Embryo
A young organism in the early stages of its development. In a chicken, this would be before hatching from the egg.

Endoparasites
Internal parasites such as worms (see page 62).

Enzyme
An agent that helps to break down food in the gut.

Evisceration
The removal of the internal organs when preparing poultry for eating.

Face
The skin around and below the eyes.

Faverolles
Breed of chicken (see page 15 for details).

FCR
Food conversion ration. This is usually expressed as the ratio of food consumed to live weight or weight of eggs produced. For example, 2 kg of feed for 1 kg of live weight is FCR 2:1 (or FCR 2).

Feather-legged
Having feathers growing down the shanks.

Feather out
When a chick loses its initial, fuzzy coat and grows its first feathers. Usually, this occurs 6–12 weeks after hatching, depending on the breed.

Feather pecking
Habit of some hens of pecking feathers from other birds (see page 67).

Fertile
Having the ability to produce a chick.

Fleas
External parasite of chickens (see page 63 for details).

Flight feathers
The primary feathers (see page 43) of the wing; sometimes used to define the primaries and the secondaries.

Flights
The main flight feathers, also called primaries.

Flock
A group of chickens living together.

Flogging the hen
An early and now archaic term for when a cock bird mounts a hen to mate with her.

Forced-air incubator
An incubator with a fan that circulates air at the correct temperature to encourage eggs to hatch.

Fowl
Collective term for chickens, ducks and geese.

Fowl Pest
Notifiable disease (see page 64).

Free-range
Describes a method of keeping chickens in which they are allowed to wander freely and to pick up casual and wild food as well as that provided for them.

French Copper Maran
Breed of chicken (see page 16 for details).

French Wheaton Maran
Breed of chicken (see page 16 for details).

Frizzle
Breed of chicken (see page 16 for details).

Gander
Male goose.

Gape worms
Internal parasite of chickens (see page 62 for details).

Gizzard
Part of the digestive system in chickens, ducks and geese where food is ground up (see page 34).

Gizzard worms
An internal parasite of chickens (see page 62 for details).

Gold Brahma
Breed of chicken (see page 17 for details).

Gold Laced Wyandotte
Breed of chicken (see page 17 for details).

Goose
Female goose.

Gosling
Young goose or gander.

Grade
To sort and classify eggs according to their interior and exterior qualities.

Grit
Very fine bits of gravel or sand; aids digestion in chickens, ducks and geese (see page 34).

Gullet or oesophagus
The digestive structure that leads from the mouth to the stomach (see page 34).

Hackle
Plumage on the rear and side of the neck of a fowl.

Hair cracks
Fine cracks on the shells of eggs.

Hamburg
Breed of chicken (see page 17 for details).

Hard feather
A classification of show birds.

Hen
Female bird that has passed through its first laying season and is more than one year old. This term applies to both chickens and ducks.

Hock
The leg joint, between the lower leg and thigh.

Horn
Describes the colour shading in the beak of some breeds, such as the Rhode Island Red.

Hybrid
The result of crossing two pure-bred lines.

Hybrid vigour
A hybrid (see above) is said to have hybrid vigour, that is displaying the best and strongest characteristics of both parents.

Immunity
Ability to resist infection.

Incubation
The process of producing chicks from fertile eggs, either naturally or artificially (see pages 54–59).

Incubator
A device that assists in the hatching of eggs (see page 55 for details).

Internal laying
A problem habit in chickens (see page 66 for details).

Isthmus
The part of the oviduct where the shell membranes are formed during the development of an egg.

Japanese
Breed of bantam (see page 26 for details).

Jumping
Term sometimes used when a cock bird mates with a hen.

Keel bone
Breast bone or sternum.

Keet
A young guinea fowl.

Lacing
A stripe or edging around a feather, differing in colour from the rest of the feather.

Layer
Term for any hen that is currently laying eggs.

Leaker
An egg that leaks because the shell is cracked and the shell membrane broken.

Leg feathers
Feathers that protrude from the outer sides of the legs of breeds such as Brahmas and Cochins.

Leghorn
Breed of chicken (see page 18 for details).

Lice
External parasite of chickens (see page 63 for details).

Light Brahma
Breed of chicken (see page 18 for details).

Light Sussex
Breed of chicken (see page 18 for details).

Litter
Material used to cover the floors of fowl houses.

Magnum
The part of the oviduct that secretes a thick albumen during the process of egg formation.

Mandible The upper or lower part of the beak.

Marking
General term for the markings on plumage, such as barring, lacing, pencilling and spangling.

Mash
Dry mash, which is milled and mixed with food ingredients and widely used to feed poultry.

Minorca
Breed of chicken (see page 19 for details).

Mites
External parasite of chickens (see page 63 for details).

Morbidity
The percentage of chickens affected by a disease.

Moult
The natural process of shedding old feathers and, later, of growing new ones.

Mounting
Term for when a cock bird mates with a hen.

Nankin
Breed of bantam (see page 27 for details).

Nematode
An internal parasitic worm that causes distress to chickens and other poultry, as well as to animals in general.

Nest egg
A wooden, china or plastic egg placed in a nest or brooding box to encourage a hen to lay eggs.

Nest run
Ungraded eggs.

New Hampshire Red
Breed of chicken (see page 19 for details).

Notifiable disease
Disease that must be reported to animal welfare authorities (see page 61).

OEGB
Term for Old English Game Bantam (breed).

Oesophagus
See Gullet.

Oil sac
Properly known as the uropygial gland, this sac is found at the base of the tail. It assists a bird when preening or conditioning its feathers.

Ovum (plural ova)
Ova are round bodies that are attached to the ovary. They drop into the oviduct and become the yolks of eggs.

Oven-ready
A bird that has been plucked, eviscerated and prepared for the oven (see pages 42–45).

Oviduct
Where eggs develop in a hen. It is formed of the funnel, magnum, isthmus, uterus and vagina.

Parasite
An organism that lives on or inside a host animal, deriving food and protection without giving anything in return.

Partridge Silkie
Breed of chicken (see page 19 for details).

Pasting
Loose droppings sticking to the vent area; also known as 'sticky bottoms' and 'pasting up'.

Pathogen
A disease-forming organism.

Pecking order
The order of seniority that naturally evolves in a group of chickens.

Pekin Bantams
Breed of true bantams (see pages 24–25 for details and their colour range).

Pendulous crop
A crop that is enlarged and hanging down in an abnormal manner. Usually found in ageing layers; it is not a serious problem.

Perch
A pole with rounded edges where hens rest and sleep.

Persistency of lay
The ability of a hen to lay eggs steadily over a long period of time.

Phenotype
The outward appearance of a bird, rather than its genetic make-up.

Pickout
Damage to the vent area as a result of cannibalism (see page 66 for details of cannibalism).

Pin feathers
Undeveloped feathers that form short stubs.

Pin holes
Holes in shells of eggs that have been caused by a bird's beak or claws.

Pinion
Tip of a bird's wing, from the last joint.

Pipping
The act of a chick breaking out of its shell.

Plumage
The feathers that make up the outer covering of fowl.

Plymouth Rock
Breed of chicken (see page 20 for details).

Point of lay (POY)
Age at which pullets start to lay eggs.

Pop hole
A doorway through which poultry are able to enter and leave a chicken house.

Poult
A young turkey.

Poultry
A generic term for species of birds that are used as food. They include chickens, guinea fowl, quails, ducks, turkeys and waterfowl.

Poussin
Small chickens, in live weight 0.5–1 kg (1–2 lb) and sold especially for the gourmet trade.

Precocity
When a pullet starts laying eggs before it is physically prepared. This usually results in small, soft-shelled eggs.

Primaries
The long, stiff flight feathers found at the outer tip of a wing (see page 43).

Prolapse
A problem habit in chickens (see page 67 for details).

Pubic bones
The thin, terminal parts of the hip bones that form part of a pelvis. They are used to judge the productivity of laying birds.

Pullet
Chicken less than one year old.

Rachis
The main and central shaft of a feather.

Range fed
Chickens that are allowed to wander and graze freely in a field.

Redcap
Breed of chicken (see page 20 for details).

Red Dorking
Breed of chicken (see page 20 for details).

Resistance
Immunity to infection.

Rhode Island Red
Breed of chicken (see page 21 for details).

Roche scale
Scale used to measure the depth of colour in a yolk.

Roo
Shortened term for a rooster.

Rooster
Male chicken (also known as a cock bird or cock).

Roosting
Describes fowl that are sleeping or at rest.

Rose comb
A style of comb, flattened to the head and covered with small nodules and finished with a leader or spike.

Rosecomb
Breed of bantam (see page 27 for details).

Roundworms
Internal parasite of chickens (see page 62 for details).

Saddle
Lower part of the back, from the centre to the tail on cockerels. In females, the same area is known as the cushion.

Scales
Small, hard, overlapping plates that cover a chicken's toes and shanks.

Scratching
The habit of chickens of digging at the ground with their claws to find insects and grains to eat.

Sebright
Breed of bantam (see page 27 for details).

Secondaries
The large, inner, wing feathers adjacent to the body; they are visible when the wing is extended.

Self-colour
A breed where the bird is of one colour throughout, such as white or buff.

Set
To keep eggs warm so that they will hatch.

Setting
Placing a group of hatching eggs in an incubator or under a hen. This is also known as 'sitting'.

Sex feather
A hackle, saddle or tail feather that is rounded in a hen but usually pointed in a cockerel.

Sex-linked
Describes any inherited factor associated with the genetics of either parent.

Sexual dimorphism
Difference between the average male and female performance within the same flock.

Shank
Leg.

Shell membranes
The two membranes that are attached to the inner egg shell. Usually, they are separated at the large end of an egg and form air cells.

Sib
Progeny of a brother-sister mating.

Sickles
The long, curved, top pair of feathers on a cock's tail.

Side-yolked
Where the yolk is offset from the egg's centre.

Silver-grey Dorking
Breed of chicken (see page 21 for details).

Silver Laced Wyandotte
Breed of chicken (see page 21 for details).

Silver Pencilled Wyandotte
Breed of chicken (see page 22 for details).

Single comb
A comb with a single, upright blade.

Sitting
See Setting.

Snood
Fleshy appendage on the head of a turkey, invariably over the beak.

Spangling
Markings produced by a large spot of colour on each feather and differing from the ground colour.

Spur
The stiff, horn-like projection found on the legs of some birds; they are on the inner sides of the shanks (legs).

Stag
British term for a male turkey.

Started pullet
A pullet that has feathered out, but has yet to start laying eggs.

Still-air incubator
Incubator that does not have the air inside it circulated by a fan.

Straight run
Chicks that are sold without first being sexed, so that male and female chicks are mixed together.

Strain
Variety of fowl that will produce the same traits when bred from one generation to another.

Tail coverts
The soft, curved feathers at the sides of the lower part of the tail.

Tail feathers
The straight and stiff feathers on the tail. In male fowl, the tail feathers are contained inside the sickles and coverts.

Tapeworms
Internal parasite of chickens (see page 62 for details).

Testes
Male sex glands.

Thigh
The part of the leg above the shank.

Thread worms
Internal parasite of chickens (see page 63 for details).

Ticks
External parasite of chickens (see page 63 for details).

Tom
Male turkey; common term in North America.

Top-knot
See Crest.

Trachea
The windpipe, part of the respiratory system that conveys air from the larynx to the bronchi and lungs.

Treading
The sexual act of a cock bird mating with a hen.

Type
The size and shape of a chicken that defines its breed.

Under colour
The colour of the downy part of the plumage.

Variety
A subdivision within a breed.

Vent
External opening through which a chicken expels both eggs and droppings.

Wattles
Thin, pendant appendages at either side of the base of the upper throat and beak, usually much larger in males than in females.

Welsummer
Breed of chicken (see page 22 for details).

White Silkie
Breed of chicken (see page 22 for details).

Yolk
The yellow part of an egg (see also Ovum).

Index